THE PORTABLE RADIO IN AMERICAN LIFE

CULTURE AND TECHNOLOGY

Series Editor
W. David Kingery

Department of Anthropology
and Department of Materials Science
and Engineering
The University of Arizona

The Portable Radio in American Life

Michael Brian Schiffer

THE UNIVERSITY OF ARIZONA PRESS

TUCSON & LONDON

The University of Arizona Press
Copyright © 1991
The Arizona Board of Regents
♾ This book is printed on acid-free, archival-quality paper.
Manufactured in the United States of America.

96 95 94 93 92 91   6 5 4 3 2 1

Library of Congress Cataloging-in-Publication Data
Schiffer, Michael B.
    The portable radio in American life / Michael Brian Schiffer.
        p.   cm.
    Includes bibliographical references and index.
    ISBN 0-8165-1259-0 (cloth) ISBN 0-8165-1284-1 (paper)
    1. Radio broadcasting—Social aspects—United States. 2. Portable
radios—Social aspects—United States. 3. Portable radios—United
States—History. I. Title.
PN1991.3.U6S35   1991
302.23′4473—dc20   91-11749
                                                             CIP

British Library Cataloguing-in-Publication Data
A catalogue record for this book is available from the British Library.

For Adam and Jeremy, and the next generation

CONTENTS

ILLUSTRATIONS

PREFACE

THIS HISTORY OF THE PORTABLE RADIO IS A WORK OF archaeology. Archaeologists study the leavings of ancient societies, from the richly appointed tombs of the Pharaohs to millions of cooking pot fragments. Mostly they deal with the latter, the common, unglamorous items of everyday existence. The portable radio in its many modern forms—boom box, Walkman, and so forth—is such an object. A necessity of life for many, portable radios are ubiquitous. Scarcely a subject for philosophic or aesthetic contemplation, its historical roots obscure, the portable radio is perfect for archaeological inquiry. When archaeologists write the history of an ancient thing, they seek to understand the contexts—technological, social, even ideological—in which that object developed over time. Thus, the common cooking pot or spear point becomes a window into the changing lifeway of a dead society.

The portable radio as surely reflects changes that took place in American society during this century. The Americas of the roaring twenties and the Depression are as dead as Rome of the Caesars. The portable radio, then, placed into those past contexts, can serve as an introduction, an opening wedge, into our own ancient lifeways. Because no item develops in isolation, the history of everyday objects is a history of the life of a people.

The archaeological approach to recent American history promises to illuminate connections that might not otherwise be suspected. For example, the emergence of the portable radio in 1939, in essentially its modern form, did not result from a technological breakthrough, but from a constellation of interconnected social and technological factors.

In addition to academics, I hope this book will appeal to those people who are ultimately responsible for its subject matter: ordinary Americans with a curiosity about how the world around them came to be.

A book of this sort could not have been completed without the contributions of many people, companies, and institutions; they deserve more thanks than I can provide here.

For furnishing information about radios, electronic history, and the electronics business, or helping me obtain the radios themselves, I am indebted to Bill Burkett, Ed Dervishian, Julius Dorfman, Francis Dukat, Roger Handy, Jack Hofeld, Matthew S. Householder, Robert E. Lozier, Randall King, Serge Krauss, Lloyd P. Morris, Phyllis Morse, Forrest Pearce, Ed and Irene Ripley, Earl Russell, Jack Saddler, Louie Schiffer, Ross Smith, Jerry Talbott, Edmund E. Taylor, Theodore Wickstrom, Michael F. Wolff, and Walter Worth; special thanks to Dave and Juanita Vaughn, Alan S. Douglas, and Harald Herp. Alan S. Douglas, who has an encyclopedic command of early radio history (and a library to match), has been extremely generous in sharing his knowledge with me. Several companies allowed me to rummage through their files, interview persons, or supplied information. I am grateful to John Taylor, John Pederson, and Carl Eilers of Zenith; Robert Galvin and Sharon Darling of Motorola; Ken and Ed Housman of Automatic Radio; Judy Anderson and Damon Davis of Howard W. Sams and Co.; O. Cooper Prude and Richard D. Hershey of Philips Consumer Electronics; and William Kendall of Arvin. The information I obtained while at Zenith was particularly important.

Greg McClure convinced me that a trip to Nebraska to study the remains of Western Manufacturing would be fruitful. I thank members of the Beshore family—Betty, Larry, Thomas, and Edward—for granting me an interview under trying circumstances. Maurice Sievers supplied information about crystal sets and commented on part of Chapter 11.

Information about subminiature tubes and their use in the first pocket radio, the Belmont Boulevard, was graciously supplied in several interviews by Norman B. Krim. Paul Carter and Kenneth M. Roemer suggested sources on portable radios in fiction. Masakazu Tani kindly translated some materials from the Japanese.

The following individuals read and commented on drafts of the book: Walt Allen, Hugh G. J. Aitken, Chuck Bollong, Alan S. Douglas, Christian E. Downum, Richard A. Gould, James Greenberg, Josiah Heyman, W. David Kingery, Norman B. Krim, George V. Leve, Randall H. McGuire, William L. Rathje, Annette Schiffer, Frances-Fera Schiffer, Louie Schiffer, Elliot Sivowitch, James M. Skibo, Masakazu Tani, Edmund E. Taylor, and Dave Vaughn. I thank them all for helpful suggestions, perceptive comments, corrections, and—above all—encouragement. Elliot Sivowitch patiently poured over the final draft and spared me from making sundry technical and historical gaffs. James M. Skibo not only worked over the first two drafts, but he also served as a sounding board for many of the ideas that strive to integrate technological and social processes. W. David Kingery gently urged me to set my sights high and dropped a few crucial hints on how to do the job better. Hugh G. J. Aitken's encouragement also meant a lot to me. Susan Douglas's *Inventing American Broadcasting*, which I read after completing the first draft, helped me to refine and extend a number of ideas and has obviously influenced Chapter 4. I thank all of these people for their significant contributions to this work.

Several institutions furnished funds that allowed me to visit libraries and archives and conduct interviews, for which I am grateful: Raytheon (former president, R. Gene Shelley) and, at the University of Arizona: Dean of Social and Behavioral Sciences, Lee Sigelman; the Department of Anthropology, William A. Longacre, Head; the Laboratory of Traditional Technology; and the Social and Behavioral Sciences Research Institute (Richard Curtis, Director).

Off-campus archival and library research was carried out at the Crerar Library, University of Chicago; New York State Library, Albany; Library of Congress; Zenith Electronics, Glenview, Illinois; the Charles River Museum of Industry, Waltham, Massachusetts; and the Archives Center, National Museum of American History, Smithsonian Institution. In every place I encountered friendly people who gave efficient and courteous service. I am especially grateful to Robert Harding, Director of the Archives Center, for his guidance in dealing with the Clark collection. Elliot N. Sivowitch, also of the National Museum of American History, showed me around the electronic collections and answered many questions. The staff of the University of Arizona libraries also facilitated my research. Although I did not visit Columbia University's Armstrong Memorial Research Foundation, the latter institution supplied useful information.

I am especially grateful to W. David Kingery and his Program on Culture, Science and Technology at the University of Arizona for a generous subvention to support publication of this book, the first in a new series, Culture and Technology.

Doris Sample, the Department of Anthropology's word-processor operator, has again done a spectacular job through three drafts of this book. J. Jefferson Reid kindly allowed me to store many radios on his shelves in the joint-use area between our offices.

The University of Arizona Press did an exemplary job in making this book; I especially thank Pat Shelton, Omega Clay, Gregory McNamee, and Stephen Cox.

Jeremy A. Schiffer and Adam J. Schiffer, my sons, cleaned many radios in preparation for photography, and Jeremy also assisted me with photography.

In July of 1988, while driving to San Diego on our family vacation, I dictated some notes to my wife, Annette. We discussed these ideas, to which she made many contributions, and they served to orient the research I undertook for the present book. When I began thinking seriously about proceeding with this project, Annette gave me strong encouragement, and her support has been unwavering despite the ever-growing collection of portable radios that has filled every available space in our home. Thank you, Annette, wife, companion, and best friend.

1

The Origin of Everyday Things

THE REFERENCE LIBRARIAN—A BRIGHT, BESPECTACLED man of about thirty—snickered in disbelief when I asked him how to look up the number of portable radios sold in 1939 and 1940. "There were no portable radios then," he snapped with authority, and added, "That was before the transistor."

Though plausible, the view that the portable radio's history is short is wrong. The real history of the portable radio, as a technological *and* commercial success, begins a half century ago, in 1939, long before the transistor had an impact on home entertainment. But, as we shall soon see, portables were manufactured much earlier. In fact, radios called by their makers and users "portable" were sold in every decade of the twentieth century. The portable radio appears even earlier in American life— as an idea in science fiction in 1890. In addition, the first completely portable radio small enough to fit in a shirt pocket contained not transistors but tiny tubes!

In America, the origin of everyday things is obscure. Few of us could answer questions about the products that surround us: the electric range, refrigerator, dishwasher, ceiling fan, television—even eyeglasses. If asked where any of these items came from, we would likely answer, not entirely in jest, "probably Japan." If pressed, we might remember when and under what circumstances each item was acquired. Thus, their history quickly becomes a microhistory, a personal history, that we recount in relation to events in our own lives: the refrigerator we bought in 1973 after our first child was born; the ceiling fan that came along with the new home, which was bought in 1979 after a big promotion; and so on.

Corporations, too, create a kind of product history—cryptohistory— that is very individualized and serves their present-day needs. Into that vast void of past time, corporations place their own accomplishments, real and imagined, to bolster their image in the eyes of consumers, employees, and investors. In advertisements, a company implies that it deserves a favored place in the consumer's heart because it has long been an innovator or was first with a product important today. Corporations also use cryptohistory in employee newsletters to rally the troops, thereby instilling pride in the company whose people for decades have made it an enlightened servant of consumer interests. Cryptohistory appears in the corporate annual report, which proclaims that a company is on the cutting edge of technology today, as it has been in the past. These statements are intended to make stockholders feel secure and to attract new ones. At its best, cryptohistory is misleading.

Unfortunately, corporate cryptohistory has a way of insinuating itself into articles and books that masquerade as real history. Uncritical authors and journalists accept corporate claims at face value, passing them along to readers who cannot judge their validity. For example, one historical "truth" reiterated in the media and popular literature is that SONY developed the shirt-pocket radio. In another example, *Business Week* in a recent (January 1990) article about Texas Instruments matter-of-factly asserted that "the semiconductor business was born in Dallas 30 years ago." Neither claim is correct. It is understandable that companies would carefully cultivate such bits of cryptohistory and disseminate them far and wide; it is less easy to understand why journalists and writers accept *and repeat* them without question.

Cryptohistory, then, is one weapon in the corporate struggle for the hearts and minds—and money—of consumers and investors. Indeed, some scholars believe that we are poised on the threshold of a new era in history, when multinational corporations become more important in world affairs (and daily life) than nation-states. Cryptohistory plays a role in transferring our allegiances—in the workplace and marketplace— from country to corporation.

Today, new products appear in stores and in the homes of friends and relatives, as if by magic. As recently as the early 1960s, most of these things were actually made in the United States. Americans of that time may not have known how or why these products were developed, or even how they worked, but at least the manufacturing process was not mysterious. After all, many people worked in factories or had acquaintances who did. In the 1990s, the manufacture of ever more household products is moving abroad, and our collective ignorance deepens. During the 1950s, about half the U.S. work force was engaged in manufacturing; today it is around 20 percent. Assembly of radios used to employ tens of thousands of Americans; now no home radios are made in this country. However, some televisions are still put together in the United States, but brand names are not a good guide. For example, SONY televisions may be made in Japan, Taiwan, or San Diego; similarly, Zenith TVs are assembled in Chicago, Mexico, and Taiwan. Not only are RCA, Philco, and Magnavox sets all made abroad, but none of these brands is U.S.-owned. The familiar things that so intimately affect our lives have become, in the last few decades, complete strangers.

How did this state of affairs arise? How could American companies, once the envy of the world in consumer electronics, have been almost completely routed? Though complex, the reasons can be investigated. As it happens, the first consumer electronic product to be swamped entirely by foreign (Japanese) competition was the portable radio, in a process that was well underway in the late 1950s. Thus, the portable radio is significant for understanding how American manufacturers came to be disadvantaged in the American marketplace. By ferreting out the history of the portable radio, we can see beyond the images created by cryptohistory that serve the interests of multinational corporations. With an understanding of what really happened in the past, informed consumers—and citizens—can scrutinize the claims and counterclaims, form independent opinions, and act accordingly.

There are several established genres of product history, each more or less faithful to what really happened in the past, yet capturing only a small part of that complex whole. One common approach is to tell the story of the inventors and of the technological innovations that propelled a product toward ever greater perfection. In this way the writers, often people with technical training only, attribute product changes to technological "progress." Products that obviously do ride on the leading edge of technology, like the airplane, tend to be treated this way. Yet, knowing that Wilbur and Orville Wright were first to solve the problems of controlled, powered flight does not help us to understand how and why the airplane became an important product that affects peoples' lives daily.

A chronology of inventors and inventions obviously captures important and often fascinating milestones in a product's history, but is woefully one-sided. Technology does not develop in a vacuum, responding only to its own internal logic and to the desires and fortunes of inventors, scientists, and engineers. The decisions to push technology and to apply it to specific products also have social and cultural dimensions. Thus, another way to write a product history is to show how social processes have set the agenda for technological progress. Such histories mostly focus on the social elite (the "taste makers"), who, in their patronage of artisans, establish trends that sometimes "trickle down" to the rest of society. Musical instruments, clothing, furniture, and fine art are products whose evolution is often treated in the context of "social history." In such accounts, which tend to be written by historians or art collectors, technological changes tend not to be discussed, even when substantial. Instead, the activities of "high society" are highlighted, and ordinary people are ignored. Yet, the activities, tastes, and needs of ordinary people—even children—also have significant impacts on technological advance and product development.

In the chapters that follow, I forge a new approach to product history that clarifies the interrelationships between technological development, the tastes of high society, and the activities of ordinary people. This approach is not actually new, for it is used daily by archaeologists. In seeking to understand the evolution of Maya temples, Hopi cooking pots, or Neanderthal spears, archaeologists exploit as much of the surviving evidence as possible, trying to illuminate the technological and social factors that each artifact reflects and embodies. And archaeologists are not elitists; they excavate trash from huts as well as from palaces. Thus, it is easy for archaeologists to appreciate that the daily activities of ordinary people also spur and influence technological development. Because the ancient inventors of long-dead societies cannot be known individually, archaeologists examine large-scale patterns of innovation. They stress not only how inventions arose but also how they came to be widely adopted.

As an archaeologist myself, I am perhaps in a privileged position to try out this new kind of product history. In effect, I will be applying an archaeological approach (behavioral archaeology in particular) to artifacts in our own society. The history of the portable radio, then, is a case study in the archaeology of us.

Of necessity, archaeology is a very object-oriented discipline, but it aims at extracting as much information as possible about the people who made and used those things found in the ground. Thus, archaeologists are accustomed to asking all kinds of questions about the technology and social processes behind each artifact. Doing the archaeology of a twentieth century object, though, is somewhat different because we can learn so much about American society and technology from historical sources. In fact, there is too much information. The problem immediately becomes that of sifting through a vast number of magazines, technical journals, and social histories to ferret out connections between that changing product and its technological and social contexts. In the process, one comes to view changes in American society from the unique vantage point of one of its products, which cannot help but be interesting. This book, then, is the history of the United States, 1890–1965, as seen by the portable radio. Perhaps only an archaeologist, equally at home in discussions of technical detail and of social change, would be foolhardy enough to tackle a product history holistically. But we really have no choice if there is to be an alternative to personal histories, narrow technological and social histories, and cryptohistories dished out by corporations.

The portable radio has a long and fascinating history, which can illustrate the general process of product development. The portable radio is ubiquitous; most families have several. No matter where one goes, portable radios will be found—on the beach, in buses and malls, even in libraries. Seemingly, one cannot escape boom box and Walkman. For many Americans, the portable radio has become a constant companion, perhaps part of their identity—at least a necessity. Because the portable radio today is nearly as mundane as a pair of socks, few people have thought about its history, its place in the lives of past Americans. It is the perfect object, utterly taken for granted, that can illuminate the processes by which our material world has come to be.

In the following pages we shall see that the portable radio is not the inexorable result of progress in electronic technology. Rather, the form that the portable radio has taken at different times is strongly influenced by social and cultural forces. The portable radio, then, is as much a product of as a product in American life.

The life of one American—Hugo Gernsback (1884–1967)—is tied intimately to the portable radio and to the story that unfolds below. According to one authority, Gernsback "is rightly known as the Father of American Science Fiction." And, if anyone deserves to be known as the father of the portable radio, it is Hugo Gernsback (Fig. 1.1). Most surprising is that the halves of Hugo—his science fiction and his radio work—are intertwined in interesting ways. Hugo Gernsback founded numerous magazines that helped to recruit American children—nearly all boys—into technical careers. In the first half of this century, when technological enthusiasm was the indomitable spirit of the times, bright

1.1   Hugo Gernsback. *Gernsback Publications, Inc.*

youngsters were fascinated by the workings of mechanical and electrical things. Turned on by visions of wondrous new products in Gernsback's science and science fiction magazines, these technological zealots became inventors, scientists, and engineers (and some even designed portable radios).

Today, more than half the students and faculty of American engineering schools are foreign nationals. Despite the immigration, an enormous shortage of engineers is expected by the year 2000. American youth are no longer attracted in large numbers to technical careers; the glitter is gone. Inventiveness in the United States—though still strong—is not what it once was. Not only is America's share of new patents declining, but our contribution of new consumer products to the world marketplace has shrunk steadily in the past few decades. The reasons for these changes, illuminated by this study of the portable radio, contribute to our understanding of America's apparent industrial decline. Many other changes took place during Hugo Gernsback's lifetime, which happens to correspond almost perfectly with the period covered by this book. In the better part of a century, the lives of Americans were transformed beyond recognition, as occupations, homes and hobbies, and the material things of everyday life all underwent dramatic changes. Once mostly frugal farmers, we have become mostly profligate urban consumers whose "needs" never end. Against that backdrop of incessant and profound change, this book explores the evolution of the portable radio.

2

Capturing Invisible Waves

IN THE WANING DAYS OF 1814, UNITED STATES TROOPS under the command of Andrew Jackson defeated the British in the Battle of New Orleans. Jackson was hailed as a hero (eventually becoming president), though this battle had no effect on the course of the war. On Christmas Eve, some days before, negotiators in Ghent—half a continent and one vast ocean away—had come to terms on a peace treaty. The war was over, but on the battlefield young men were still fighting—and dying. Tragedies like the Battle of New Orleans were commonplace before the advent of the telegraph, telephone, and radio, which provided "real time" communication.

America in Andrew Jackson's time—he was president from 1829 to 1837—was largely a land of farmers and small merchants. As late as 1850 only one-fifth of the population lived in towns larger than 2,500 people. However, based on technology appropriated illegally from the British, the industrial revolution had begun in the United States. In the Northeast, cities sprang up around the textile mills, the first driven by water power, later ones by steam engine. Labor shortages, along with an abundance of raw materials, spurred an inventive spirit unmatched in human history. Americans, rural and urban, had an abiding faith that the ideals of their new democracy could be realized through labor-saving inventions, which would make food and manufactured goods plentiful and accessible to all.

By 1850, when the United States had, through conquest and purchase, extended its boundaries from the Atlantic to the Pacific, problems of transportation and communication loomed large. Exploiting domestic and foreign inventions (and often utilizing immigrant engineers and laborers), U.S. entrepreneurs responded with canals, railroads, steamships, and the telegraph. These technologies permitted government control to extend gradually over the new territories, enabled manufacturers to dispatch their goods to merchants they had never met, and hastened the Americanization of disparate peoples. Taking advantage of these transportation improvements, the postal service allowed Americans to share their innermost thoughts with others across the continent.

Communication, however, was far from reliable or rapid. A letter to Tucson was as likely to be intercepted by bandits or Apaches as to reach its destination; and the telegraph—based on Morse code—was hardly suited for daily business or personal use. To send a telegram, one had to visit a Western Union office, which was not always convenient.

The coming of the telephone, in the late 1870s, promised—and decades later delivered—a communications revolution. For the first time, supply and demand could be matched by business instantaneously. Both telegraph and telephone, however, were tethered to poles and wires. Though cables could be laid across oceans and lakes, the expense was great and was recouped by high transmission fees. Over land, "long-distance" meant hundreds, not yet thousands, of miles, until the early twentieth century—and it was expensive. Nonetheless, the telegraph and telephone contributed to (and reflected) the heightened pace of industrialization that began in the last decades of the nineteenth century.

The new industries were unique in their dependence on technical knowledge. Scientific disciplines like chemistry and physics were presented with challenging problems from industry and flourished as never before; entirely new areas were opened to inquiry. A few scientists and engineers even began to find work in corporations. Other people with technical training, sometimes self-taught, became independent inventors, from whose workshops and laboratories flowed a constant stream of novel ideas and products. Thomas A. Edison was the best-known independent inventor, but was hardly typical; his large laboratory was established to be nothing less than an invention factory. Well-equipped and well-staffed, Edison's facility gave the world a succession of important products, beginning in 1879 with America's incandescent lamp—better known as the light bulb.

Although limits to the expansion of telegraph and telephone service were obvious by the 1880s—they could never encompass every remote outpost or ships at sea—neither AT&T nor Western Union, the monopoly companies, invested much money in developing new technologies. To extend the existing systems would require a true breakthrough—wireless communication. The telephone company carried out research, to be sure, but only that concerned with improving the existing art. Thomas Edison did desultory work on "wireless" in the early 1890s, but it bore no fruit. For the origins of wireless, one must look elsewhere, across the Atlantic.

In December of 1901 the Atlantic was for the first time spanned by a wireless message at the speed of light—186,000 miles per second. In Newfoundland Guglielmo Marconi had received the letter S in Morse code from his transmitter in Poldhu, England. Although the Atlantic can be crossed today with a shirt-pocket transceiver, Marconi's apparatus was somewhat less portable. In his marvelous book on early radio, *Syntony and Spark*, Hugh Aitken described the enormous transmitting antenna Marconi erected at Poldhu: "A ring of 20 wooden masts, each 200 feet high, arranged in a circle 200 feet in diameter. From the tops of these masts there was suspended a conical arrangement of wires gathered together at the lower point in the shape of a funnel." Though not surviving the first major storm, this latter-day woodhenge and its replacements established the feasibility of intercontinental wireless communication.

Wireless did not spring full blown from Marconi's mind, fertile though it was. Rather, Marconi was the first to improve upon and commercialize discoveries made by some of the nineteenth century's greatest physicists. Marconi appreciated the problems of long-distance communication that only wireless could solve, and he exploited physics for a practical solution.

Wireless transmission (called "radio" after about 1915) depends upon the existence of invisible electromagnetic waves, which were first predicted by James Clerk Maxwell. Maxwell is unknown to most people, but among physicists he has the same stature as Newton and Einstein. Caltech's Richard Feynman, himself a noted physicist, has said that "the most significant event of the nineteenth century will be judged as Maxwell's discovery of the laws of electrodynamics." These laws describe the behavior of the electromagnetic waves that make possible radio, television, radar, and most other advanced technologies we enjoy today.

Maxwell was born in 1831, in Scotland, to a family of the landed gentry. As a child, he had a keen—even demanding—curiosity into the things and phenomena of the natural world. Despite having for a time an unsympathetic tutor who beat him over the head with a ruler and twisted his ears until they bled, Maxwell published his first scientific paper at the age of fourteen. Because of his considerable mathematical gifts, he was sent to Cambridge University. He graduated in 1854, and then held an assortment of academic posts in England and Scotland. An utterly ineffectual lecturer, Maxwell left most of his students bored, lost, or both. However, his early researches on color vision—which included a demonstration of the first color photograph in 1861—and on the rings of Saturn established him as a preeminent physicist. Like many academic heavyweights, Maxwell received the call to take on administrative duties. Funds to build a physics laboratory had been donated in 1870 to Cambridge University by the Seventh Duke of Devonshire, William Cavendish, who was also the university's chancellor. Maxwell was awarded Cambridge's first chair of experimental physics and designed the new laboratory, which opened officially in 1874. The Cavendish Laboratory quickly became a world center for experimental physics and modernized the teaching of science at Cambridge.

Maxwell's most significant work, which led to radio, was on electromagnetism. Michael Faraday had shown in the 1830s that a new conception of electricity and magnetism was sorely needed. Electricity could be turned into magnetism and vice versa, but the old Newtonian models explained these effects only clumsily. Faraday introduced the idea of a *field*, suggesting, for example, that a compass is deflected by a current-carrying wire because of the latter's magnetic field: the flow of electricity actually alters the space around the wire. Similarly, a moving magnet could, also by virtue of a magnetic field, induce a current in a nearby coil of wire. Faraday's idea of field is pivotal in modern physics, but it did not become widely accepted until transformed into a full-blown mathematical theory by Maxwell in the early 1860s, a few years after the first telegraph message was sent across the Atlantic on a submarine cable.

Like the waves of sound in the air and waves of water on a pond, electromagnetic waves, it was thought, had to be propagated through a medium—the "luminescent aether" or, simply, the ether. Describing the properties of that intangible medium occupied Maxwell in the 1870s. Only much later was the puzzle of the ether solved by Einstein.

Maxwell's equations predicted the existence of invisible electromagnetic waves similar to both light and magnetism. Although Maxwell was aware of the implications of his equations and was himself a skilled experimenter, he did not attempt to create and capture invisible waves in the laboratory. Maxwell was quite content to build his theoretical edifice and work on other challenging problems—like the nature of the ether. It is doubtful, though, that anyone in the 1860s or 1870s attached practical importance to invisible waves or had envisioned their use for communication.

Nonetheless, invisible waves were important; if they existed as predicted, then Maxwell's theory would be given enormous support. Typically, Maxwell did not assign students to work on his pet projects (a reticence that even then was quaint), so no one at the Cavendish pursued invisible waves. Not until his two-volume work, *Electricity and Magnetism,* was published in 1872 did Maxwell's revolutionary theory become widely known among physicists. Still, there was no stampede to test it.

While physicists were plodding through the equations in *Electricity and Magnetism*, a more popular literature was bringing to Americans and Western Europeans visions of exciting futures based on new technologies. In 1843, for example, William Samuel Henson, a lace trader in Somerset, England, published the world's first design for a fixed-wing aircraft driven by a propeller. The lovely engravings of Henson's "aerial steam carriage" were disseminated widely in books and articles. This vision achieved great popularity and influence throughout the literate world.

Visions like the "aerial steam carriage," even when fanciful and utterly impractical at the time, can play an important role in technological development. When a vision is sufficiently popularized, it may attract a group of adherents—a constituency—and become a cultural imperative. A cultural imperative is a product believed by its constituency to be desirable and inevitable, merely awaiting technical means for its realization. The constituency of a cultural imperative, which may be a rather small but vigorous group, often takes concrete steps toward creating the necessary technology. Cultural imperatives, then, furnish mandates for technological development.

Visions that may become cultural imperatives often originate in popular literature. For example, the writings of Jules Verne, beginning in 1863, helped to create sizable constituencies for the cultural imperatives of air and space travel. The Verne novels, notes Charles Gibbs-Smith in *Aviation*, dealt with "aeronautical romance, speculation, fantasy, and prophecy, which undoubtedly played a large role in firing the imagination, and acclimatizing the minds, of Victorian youth around the world." One of these Victorian youths was an American, Robert Goddard, who read Verne's *From the Earth to the Moon*. Later, as America's most distinguished rocket scientist, Goddard acknowledged his debt to Verne's inspiration.

Thus, imaginings of the future are signposts that can orient people's technological activities. Humans, we may be sure, are the only animals who spend time pursuing visions of things—and a world—that might be. As philosopher Ernst Cassirer has noted in *An Essay on Man*, "To think of the future and to live in the future is a necessary part of man's nature."

When a vision has become a cultural imperative, efforts dedicated toward its realization—by inspired individuals, companies, even governments—are regarded as risks worth taking. By the late nineteenth century, for example, tinkering with gliders, airships, and airplanes had become an activity that made sense to many people. Cultural imperatives, then, stimulate experiments that apply existing technology and at times lead to the creation of new technology. When a cultural imperative is seen to have military value, for example, governments will try to hasten the pace of development, sometimes through massive investment.

The concept of cultural imperative helps us to understand efforts undertaken to create a product, but the factors that lead to its purchase by consumers are quite different. Indeed, even if a cultural imperative has a large constituency and a practical product is actually produced, it may still fail in the marketplace. The picturephone, for example, has been a cultural imperative since the late nineteenth century. However, when practical technology was introduced into a test market in the early 1970s, the picturephone flopped.

One cultural imperative that took hold in the late nineteenth century was the wireless telegraph. Popularized in science fiction, this vision was also clearly brought into being by real communication problems. The technological basis for creating wireless lay in Maxwell's invisible waves. But, first, there was a need for more basic physics.

The challenge of invisible waves was finally taken up by the German physicist Heinrich Hertz in the late 1880s. Hertz presumed that if Maxwell's theory were correct, he should be able to create and capture such waves in the laboratory. A brilliant experimenter, Hertz built the necessary apparatus.

Like all radio systems since then, it consisted of a sending unit (transmitter) and a receiver (Fig. 2.1). Using a spark coil, the transmitter created an alternating flow of electricity in a circuit: first it went in one direction, then in the other, reversing many millions of times per second. This oscillating circuit generated a radio wave that could be broadcast by means of a wire antenna. The number of oscillations per second is known as the radio wave's frequency and is expressed today in units of Hertz (for example, 980 kilohertz). To pick up a particular transmission, the receiver is tuned to that frequency with an antenna of proper length and appropriate components (capacitors and inductances) in a tuning circuit. Hertz's receiver was simply a wire ring. Nonetheless, when the transmitter was turned on, a tiny spark jumped between the open ends of the receiving wire, confirming that invisible waves had crossed the room.

To test Maxwell's theory required that Hertz be able to measure independently the length of the radio wave and its frequency; from these quantities, its velocity could be calculated. In 1887 he succeeded in showing that electromagnetic waves travel at the speed of light. Maxwell's theory was irrefutably supported.

Hertz himself did not file patent applications and form a radio corporation; his passion was physics, to understand a little more about the innermost workings of the world. Hertz, however, did not toil in obscurity, for he published his results rapidly in physics journals. As a result, his experimental wizardry became renowned in the scientific community and even beyond as this major discovery was popularized in magazines and newspapers. In addition, invisible waves were made to perform countless times before delighted audiences in lectures at universities and scientific societies.

In the late 1800s, while European scientists like Hertz were building the foundations of radio, Americans—especially novelists—were divining a better world. The capitalist system seemed on the verge of collapse, for the gulf between rich and poor had become huge during the period of rapid industrialization that followed the Civil War. Economic power was concentrated in the hands of the "robber barons" and trusts, while vast armies that included women and children toiled long hours in dismal factories. A great many people, especially recent immigrants, were ill-clothed, ill-housed, and ill-fed. Angry workers, stung by the Panic of 1893, were ready for radical action. These problems, as well as the crime, disease, and squalor of eastern cities, set a generation of reform-minded writers to work. Surely a more humane industrial society could be envisioned.

2.1 Heinrich Hertz and his experimental radio apparatus, 1887

The utopian novels of the 1880s and 1890s presented bold blueprints for the future. Though most of these books focused on imaginative social arrangements, such as World government and a moneyless economy, developments in technology were not completely ignored. In Chauncey Thomas' *The Crystal Button*, of 1891, the inhabitants of the city of Tone—Boston in 4872—enjoyed airships and monorail trains that traveled 200 mph.

In *A.D. 2000*, author Alvarado Fuller prognosticated solar power, instant cameras, and contact lenses. That same book of 1890 also described the discovery of the North Pole by aeronauts. While at their icy destination the explorers communicated with the president of the United States. This long-distance call was made possible by a large box that one man could carry from place to place. Called a "Sympathetic telegraphic system," it was, in all respects, a portable radio. In science fiction, at least, the portable radio had entered American life.

Among the other technological achievements forecast by the utopian novelists were robots, picturephones, color television, and breath deodorizers.

The most popular book of this genre, and one of the best selling novels of the entire nineteenth century, was Edward Bellamy's *Looking Backward*, published in 1887—the same year as Hertz's momentous experiment. *Looking Backward* projected a harmonious new society in America of 2000, with universal college education but no bankers or lawyers. Among the technological marvels of that future age would be electricity for heating and lighting, credit cards for every purchase, and large pneumatic tubes for distributing goods from warehouse to home. Of most interest, however, is the home music room. After consulting a program that listed the day's offerings, the listener adjusted "one or two screws," which filled the room with music "perfectly rendered." The program came to every home via telephone from central music halls where the best musicians performed twenty-four hours a day. On Sunday mornings, there was even a choice of sermons.

American homes of the late nineteenth century were by no means without music. A few had crank-operated cylinder phonographs, but they were expensive and sounded pretty awful. Wealthy families could also afford pianos, which cost several hundred dollars. The less well-to-do might buy a parlor organ, an imposing bit of indoor architecture, which sold for a still-ghastly $25 to $400. Played by the wife, the parlor organ was almost a necessity for the Victorian middle-class family; doubtless many scrimped and saved to get one. In its own way, the organ made the parlor—wall-to-wall with throw rugs and lush wallpaper—in Victorian homes into a music room, but one requiring constant participation. Needless to say, Mom was not always free to make music. Even the phonograph needed frequent attention, for a cylinder lasted only a few minutes and the spring-driven mechanism needed periodic cranking. The passive music room of Bellamy was a striking image.

Live concerts, of course, were mainly social affairs for the rich. Bellamy had created a society in which everyone—not just the wealthy—could hear in their own homes the greatest musicians and singers. The democratization of music was a remarkably prescient vision—and so typically American. A music room could provide the most humble worker with the same pleasures heretofore enjoyed exclusively by the robber barons and their snooty ilk. The idea of the music room, like airships and electric heating, entered the American consciousness, a cultural imperative awaiting a means of realization.

As an extension of telephone technology, the music room was not far-fetched—even in 1887. Prior to then, Alexander Bell himself had employed music to demonstrate the telephone. By connecting a large metal funnel or horn to the earpiece of a telephone, the sound could be amplified sufficiently to be heard by people in the room. This "loud-speaking" telephone, which would become important as the horn speaker in radio, early on opened the door to the possibility of using the telephone for entertainment broadcasting. Although successful demonstrations were carried out, the telephone company was not interested in this application of their technology. The music room would have to await the *wireless* telephone based on Maxwell's invisible waves.

It is doubtful that many of the utopian novelists were familiar with invisible waves, for most lacked technical training. Bellamy, for example, was a lawyer turned journalist. However, Fuller, who envisioned the portable radio, was a lieutenant in the United States Army and had a keen interest in communications. In any event, it would take visionaries of a different sort to see in Hertz's experiment the seeds of a worldwide communication network. A few scientists, including Elihu Thompson and Sir William Crookes, did immediately foresee the possibility of a wireless telegraph, but they did not embark on a developmental effort. It was Guglielmo Marconi who created the first practical wireless system.

Of Italian father and Irish mother, Marconi was a directionless youth when he learned about Hertz's experiments and invisible waves from his Bologna neighbor—and sometime mentor—Augusto Righi, himself a physicist. At once he was consumed with a passion to transform this wonderful discovery into a practical communication system that could enrich mankind—and, perhaps, Marconi. At his father's villa, he tinkered incessantly with wireless apparatus, making minor improvements that gradually extended its range.

A new antenna design in particular had enabled Marconi to get gains in distance. The antennas (for receiving and transmitting) were erected vertically, suspended from tall poles. The antenna circuit also included a grounded connection, in which a metal plate was actually buried in the earth. (Ground connections would still be used on some home radios well into the 1930s.)

Marconi first offered his improved radio system to the Italian government. Rebuffed by the Italians, the persistent Marconi took his inventions to England in 1896 where the reception was somewhat warmer (Fig. 2.2). England at that time was the hub of a global empire ("from Aden to Zanzibar") on which the sun never set. Improving communication with far-flung colonies had been a constant concern, a deeply ingrained cultural imperative. Indeed, when Marconi arrived in England, experiments with other kinds of "wireless telegraphy" had been underway for many decades. For example, using magnetic induction, Sir

2.2 Guglielmo Marconi and the wireless system he brought to England in 1896

William Preece in 1892 had sent messages across the Bristol Channel. However, magnetic induction was never to become practical for great distances because the antennas (both transmitting and receiving) had to be as long as the distance to be spanned. As Hugh Aitken explains in *Syntony and Spark*, this technology had reached a dead end and that is why Preece himself, a government official in the post office, took Marconi's research seriously and became his champion.

Marconi applied for a patent on his apparatus and in short order had, with the prodding and financial backing of his mother's family, established the first Marconi radio company. A subsidiary was soon set up in the United States.

The earliest commercial deal of any consequence was with Lloyd's of London, the venerable insurance syndicate. Receiving and sending devices were installed aboard ships to improve the safety of navigation in dangerous waters and in bad weather. By the first year of the new century, Marconi companies had established ship-to-shore communication as the first truly practical application of wireless telegraphy.

Though actual uses of wireless were still limited, information about the new technology was spread widely in technical journals and popular magazines. Both its potentials and drawbacks were explored, as people groped to fathom the new wonder. Wireless communication was regarded, appropriately enough, as something quite remarkable, perhaps destined to replace the telegraph and telephone. In mentioning Marconi's work, *Scientific American* gushed in 1897 that "this system of telegraphy and signaling has capabilities within it that will astonish the world." Further, the article asked, "Will it enable a person at one point on the globe to communicate with another on the opposite side? Can this energy be utilized in communicating with other worlds?" In most articles, potential uses of the new technology were forecast, including military communication in the field and ship-to-ship communication. An article in *Harper's Weekly* in October 1899, even suggested that wireless would make possible communication with balloons and trains.

Prior to the end of the century, however, virtually all of wireless's anticipated uses involved point-to-point communication; that is, telegraphy and telephony without wires. Indeed, the limitations of the emerging technology were framed precisely with those applications in mind. The problem with wireless communication, many experts opined, was that it would not be private. Typical was this 1899 editorial in *Scientific American*:

> Before wireless telegraphy can be used for general commercial purposes some method must be devised whereby, as in wire telegraphy, the transmitter can communicate with one particular receiver to the exclusion of all others, and the receiver can exclude all messages except the particular one directed to it. Until this is achieved the new system must be barred from the field of ordinary commercial work.

To solve the broadcasting "problem," Marconi and other inventors developed ever more selective tuning circuits. It seems remarkable today that wireless technology was so thoroughly regarded as a mere extension of the telegraph and telephone. Not until after the turn of the century would visionaries see in the new technology entirely novel possibilities. It is common that, in actual trials of new technologies and products, new applications emerge; sometimes the latter turn out to be the most important ones, though unforeseen by earlier visionaries and the inventors themselves.

The apparatus that Marconi brought to England in 1896 was, in a sense, a portable wireless system (Fig. 2.2)—small and easily carried. The apparatus he used in his first public lecture in London consisted of two small boxes placed some distance apart in the hall; when the transmitter was turned on, a bell sounded in the receiver. As Marconi's experiments continued in England, more powerful transmitters were built to obtain greater distances with the lower frequencies (thus longer waves) being used at that time. (It was believed, erroneously, that short waves did not travel very far.) Thus, as transmitters became more powerful, they grew to enormous size—often filling a room. Even receivers lost portability as the number of components needed for tuning—capacitors, inductors, and so forth (which then were quite large)—multiplied.

Although much practical wireless apparatus was losing rather than gaining portability, some people did envision a more personal wireless that could be easily moved about. These visions, however, were often expressed skeptically. Writing in *Electrical World*, Arthur R.V. Abbott offered in 1899 the following judgment: "Telegraphing without wires— how attractive it sounds. . . . A little instrument that one can almost carry in the pocket. Possible? Certainly, but will it pay? For this is the final criterion with which this utilitarian age tests all such propositions, and for the present . . . the answer must be NO." Likewise, an author in *The Living Age*, also writing in 1899, mocked one Herr Schaefer, who took credit for inventing a "new system of wireless telegraphy." Using this system, said Herr Schaefer, "two persons long distances apart, provided they each have my little apparatus, can converse just as easily and distinctly as with our well-known system of wire telephones." To this claim the author replied, "The 'little apparatus' is a thing apparently to

carry about in the waistcoat pocket, and no doubt there are 'full direc-
tions' with each instrument supplied. Seriously, the world is moving too
fast for sober people." Evidently, the idea of a pocket wireless was still
too outlandish to attract many adherents. In just a few years, however,
children would actually be building them.

The most portable wireless receiver of this time was used in experiments by one Dr. I. Kitsee, which he reported in *Electrical World* in
1899. The apparatus was little more than "two ear receivers with an
adjustable band"; the wearer's body was the antenna, and a metallic plate
on one shoe the ground. When near a sending station, the receivers simply clicked—this was no-frills wireless of near ultimate portability.

In the final years of the nineteenth century, the United States was again
at war. The Spanish-American War above all testified to the growing
influence of newspapers. Their expansion during the previous decades
had been phenomenal: in 1880, 850 daily papers had a total circulation
of 3 million; by 1910, it was 2205 papers with a circulation of 22.4 million. In a world that became more complex and bewildering day by day,
where constant change was almost commonplace, newspapers and magazines provided a smorgasbord of information—on national and world
events, on new inventions, on the social scene, on sports, and so on—
perhaps helping the reader to make some sense of what was going on.
Newspapers and other print media shaped new American attitudes and
values, as they chronicled—and cushioned the impacts of—change.

The sinking of the battleship *Maine* in Havana harbor was the kind of
event whose meanings could be readily manipulated by the press. Two
hundred and sixty Americans died on the *Maine*; the newspapers reported that Spain, directly or indirectly, had to be responsible. In the
face of an aroused public, President McKinley, fearing for his political
life, called on Congress to declare war on Spain. Congress responded
overwhelmingly, and McKinley signed the war resolution on April 20,
1898. The United States would have the opportunity to put its new all-
metal navy to the test.

Despite the rapid progress in wireless, when Commodore Dewey defeated the Spanish fleet in Manila Bay, news of that event—traveling by
ship and cable—took a week to reach the United States.

In the new century, wireless would play a large role in war: in readying
a people to accept its inevitability, in coordinating military activities, in
providing a basis for new weapons, and in reporting its horrors and
heroes back home. By the same token, war would play a large role in the
development of wireless. Though military activities were an incentive to
push electronic technology to new heights, corporate and other civilian
activities and interests also provided, at times, a powerful impetus to
radio. It is the interplay of these diverse forces in shaping radio, and
portable radios in particular, that underpins the story to be told in the
following pages.

3

Wireless Technology Comes of Age

JUST BEFORE THE TURN OF THE CENTURY, A YUGOSLA-vian immigrant, Nikola Tesla (Fig. 3.1), went west to Colorado to carry out bizarre electrical experiments with high voltage and high frequency alternating current. Tesla created artificial lightening as well as other electrical phenomena never before seen. After his return, he expounded new and breathtaking visions for wireless.

Tesla was one of the first modern techno-mancers—someone with technical training, or knowledge of technological progress, who looks deeply into the current art and finds indications of future developments. In ancient China, cracks on roasted shoulder blades of deer foretold the future (an example of scapula-mancy). Most cultures, in fact, found clues to forthcoming events in the stars, in trances and dreams, and even in chicken entrails. In twentieth-century America, the future has been sought in science and technology. Like the oracles of ancient Greece and Rome, eminent scientists and inventors such as Tesla have been consulted about potential new products. Their pronouncements in the media give rise to—and give voice to—cultural imperatives.

With the arrival of the new century, "technological forecasting" became a cottage industry. Dozens of articles, most with "future" in their titles, appeared in every conceivable place, from *Cosmopolitan* to *Ladies Home Journal*. Even the *Annual Report of the Smithsonian Institution* carried technical prophecies. Techno-mancy was fast becoming techno-mania.

Tesla was an independent inventor, a figure already famous in the electrical industry. It was his designs for alternating-current generators and motors that made possible the modern electrical power system. In "the war of the currents" waged by George Westinghouse (alternating current or AC) and Thomas Edison (direct current or DC), Tesla's inventions finally won out. With alternating current, power could be transmitted cheaply over long distances and put to work in factories, businesses, and homes. The first major installation of the Westinghouse-Tesla polyphase system was at Telluride, Colorado, in 1891. That same year the White House was also electrified.

3.1   Nikola Tesla

While Westinghouse was doing battle with Edison to electrify America, Tesla's interests turned to high-frequency alternating currents. The latter currents, of course, create electromagnetic waves—the basis of wireless. Tesla was no stranger to wireless. In the early 1890s, not long after Hertz's work was reported, Tesla initiated investigations into wireless and anticipated several of Marconi's crucial inventions. In fact, the U.S. Supreme Court upheld the priority of Tesla and others over Marconi in important radio patents in 1943—the year Tesla died nearly penniless. (Unfortunately, Tesla's inventions never quite provided enough money to support his ambitious experiments, and so he struggled constantly to maintain a flow of funds from wealthy benefactors.) Tesla's laboratory burned down in 1895, with the result that much valuable information, wireless apparatus, and time were lost. Nonetheless, even by the end of the decade, Tesla's ability to command invisible waves was second to none; Marconi himself watched his progress nervously and was thereby spurred to be first to communicate across the Atlantic.

In contrast to Marconi, Tesla had little interest in commercializing his wireless inventions. Rather, he employed them in startling displays to attract research funds. Typical was Tesla's contribution to the Electrical Exhibition of 1898 at Madison Square Garden: a submersible model boat, controlled entirely by wireless. At Tesla's command, the battery-powered boat moved in any direction. This was a remarkable feat, embodying uses for radio that would not become important, or even fully practical, for decades. However, in 1898 few people took notice. Tesla's visions, for example, of robots and guided missiles controlled by wireless and death rays, sometimes seemed outlandish, even fanciful, when originally proposed. Increasingly, other engineers and scientists distanced themselves from Tesla's imaginative projections, especially after he claimed reception of wireless signals from Mars.

The experiments in Colorado, undertaken after the Madison Square Garden exhibition, went well, though the eccentric inventor also managed to burn out the local power company's generator, which he and his assistants quickly rebuilt. Tesla returned to the East with an unsurpassed appreciation of the extraordinary properties of high-voltage, high-frequency electricity. The Colorado experiments had convinced Tesla that it would be possible to transmit, *without wires*, not only information over long distances but also electric power. Large central stations would radiate energy, and homes and businesses with suitably equipped receivers could tap off the power they needed. In 1900 he published a long article in *Century* magazine that included striking photographs from the Colorado experiments as well as discussions of the potential for long-distance transmission of intelligence and electrical power. The article kept Tesla, the magician, in the public eye, which is precisely what he needed in the constant quest for cash. These visions were grandiose, even for Tesla, but he moved forward to demonstrate their feasibility.

With J. P. Morgan as backer, Tesla began building a monstrous transmitting tower at Shoreham, Long Island. It was 187 feet tall, an ungainly derrick capped by a hemisphere, and was to be the centerpiece of the "world system of communication" (Fig. 3.2). In several brochures intended to attract more financing, Tesla expounded his stirring vision. In addition, his 1904 article in *Electrical World* made clear that wireless's supposedly grave defect—indiscriminate broadcasting—was actually its greatest virtue. He also furnished a mandate for the pocket portable radio, which soon would become a cultural imperative. In his own words, the world system

> will prove very efficient in enlightening the masses, particularly in still uncivilized countries and less accessible regions. . . . A cheap and simple device, which might be carried in one's pocket may then be set up anywhere on sea or land, and it will record the world's news or special messages as may be intended for it. Thus the entire earth will be converted into a huge brain, capable of response in every one of its parts.

Regrettably, Tesla supplied no details of the pocket wireless receiver. Sadder still, he was unable to raise enough money to complete the transmitting station. The unfinished hulk endured until late 1917, when the U.S. government, fearing the tower was being used by spies, dynamited it.

Tesla, already bitter that Marconi—infringing his patents—had sent the first wireless message across the Atlantic, never recovered fully from this failure to build the world system. Some years later, J. Stone Stone, also a distinguished wireless pioneer, said that Tesla "was so far ahead of his time that the best of us then mistook him for a dreamer."

Tesla's visions not only outran his money-raising skills, but also wireless technology. Before world systems and pocket radios could become possible, wireless technology would have to be pushed along some distance. Fortunately, Tesla was not the only dreamer at work on wireless at the beginning of the new century. Other independent inventors took up the challenge; they did make most of the major technological breakthroughs, but often lacked resources to perfect and effectively market their inventions. As commercial and national defense applications of wireless became evident, large corporations and the military assumed an important role.

At the turn of the century, the detector was the weakest link in the receiver. (The detector separates the signal from the high-frequency carrier wave by allowing electrons to pass in one direction only.) Various devices had been employed as a detector, including the "Branly coherer," a glass tube containing iron filings. It was, in fact, Marconi's improvements that made the Branly coherer into a practical detector, though it had to be reset—by tapping—after each dot or dash. This cumbersome feature severely limited the speed at which messages could be sent. Also, mechanical coherers could not detect signals other than Morse code, a disadvantage that was decisive later when voice and music wafted through the ether.

Thus, within the fraternity of wireless engineers and electrical experimenters, a cheap, efficient, and reliable detector had become an insistent cultural imperative. Many people worked on the problem, and it was quickly solved. Several solutions led to inventions that were nothing less than momentous: the vacuum tube and the crystal diode. The former would usher in the age of electronics and the latter—much later—the solid-state revolution (see Chapter 12).

The crystal diode was established as an important detector in the period from 1902 to 1905 by Greenleaf Whittier Pickard. In an extensive series of experiments at AT&T's Boston laboratory, Pickard found that certain mineral crystals like galena and iron pyrites made excellent detectors. Though cheap and efficient, crystal detectors were not without some drawbacks. In order to get detection, a small wire (the "cat's whisker") had to be precisely positioned on a sensitive spot on the crystal's surface. Although requiring some set-up time and periodic readjustment, crystals rapidly became the detector of choice for most applications, and remained so until the twenties.

The origin of the vacuum tube detector actually begins with Thomas Edison and his light bulb. In trying to understand why the interior of light bulbs darkened with use, Edison found in 1883 that a hot filament gives off electrons. (Actually electrons were not discovered until 1897, but I will describe his findings in modern terms.) To study this emission in more detail, he placed a small metal plate in the bulb at some distance from the filament, connecting the plate to the positive terminal of a battery and the filament to the negative. Electrons (which are negatively charged) were attracted to the positive plate, and so a circuit was created involving a unidirectional motion of electrons (from filament to plate). This phenomenon we know today, appropriately enough, as the Edison Effect. As was typical for Edison, he thought up a new circuit that included the light-bulb-with-a-plate (it was trivial and never used) and was awarded a patent. Edison did no further experiments along these lines, moving on to more lucrative inventions such as the phonograph and motion pictures.

A year after discovering the one-way flow of electrons in a light-bulb-with-a-plate, Edison was visited by a young British electrical engineer, J. Ambrose Fleming, for whom he demonstrated the effect. The training of this engineer, by the way, had included a stint at the Cavendish laboratory. Fleming eventually went to work for Marconi and designed the transmitter at Poldhu that first broke radio silence across the Atlantic. A few years later, in 1904, Fleming put the Edison effect to work in wireless. He appreciated that a device allowing electrons to flow in but one direction could be used as a detector; Fleming's new detector, then, was simply a light-bulb-with-a-plate—a two-element vacuum tube (called a

3.2   Tesla's "World System of Communication." *Left,* the Wireless Tower; *right,* artist's conception of the World System in operation

diode). Although Marconi's ship installations employed "Fleming valves" (as they were called in England), it would be some time before tubes were used as detectors routinely (Fig. 3.3).

In the final analysis, the Fleming valve was less significant as a detector than as a bridge to an entirely novel vacuum tube, the audion, which amplified the feeble sound. Lee de Forest created the audion, a bulb with a very bright future.

De Forest, from a family of modest means, was a colorful character with an unbridled ambition to become rich. However, lack of business acumen and a propensity to become involved with sleazy promoters thwarted his ambition for some years. He was, however, a first-rate inventor, perhaps even deserving to be called the father of radio, which is the title of his autobiography.

Like Marconi and Tesla, de Forest appreciated the importance of technological display—staging public events to advertise his progress in wireless, which could stimulate the sale of products and stock in his companies. Toward these ends, de Forest put wireless apparatus in then-unusual places. Among these public "feats" were communication with a hot-air balloon and the first auto radio in 1904 (Fig. 3.4) during the St. Louis International Exhibition (where the modern hot dog was also introduced). In the following year, de Forest paraded his wireless car, including a brass pole antenna and ground wires dangling from its rear, along Michigan Avenue in Chicago. A message was picked up from de Forest's transmitter in the Railway Exchange Building. De Forest's partner was quoted in the *Chicago Tribune* as follows: "We hope it will be possible for business men, even while automobiling, to be kept in constant touch with LaSalle Street." These displays served de Forest's pecuniary interests, to be sure, but they also expanded the scope of potential wireless applications, helping to create new cultural imperatives. "Mobile" wireless, heretofore restricted to ships, now included that in cars and aircraft.

In the same year, 1905, that de Forest demonstrated the auto wireless in Chicago, he learned about the Fleming valve. Having already experimented with detectors, de Forest took a keen interest in the new invention; this he later denied. However, Gerald F. J. Tyne has shown that late in 1905, de Forest commissioned H. W. McCandless, a maker of light bulbs, to build several Fleming valves. In experiments with these custom-built tubes, de Forest came up with the idea of adding a third

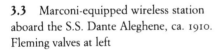

3.3 Marconi-equipped wireless station aboard the S.S. Dante Aleghene, ca. 1910. Fleming valves at left

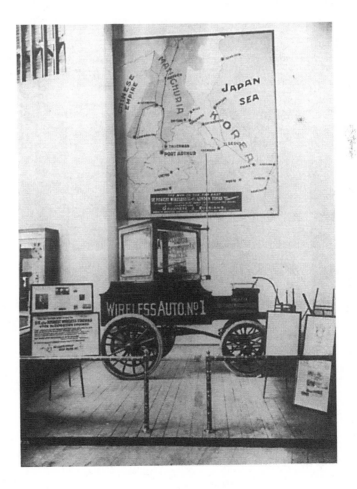

3.4 Lee de Forest's "Wireless Auto, No. 1" at the St. Louis Exhibition, 1904

element (creating a "triode"). Between the filament and the plate he inserted a small piece of metal, which resembled a gridiron. A negative voltage applied to the "grid" (the name stuck) interrupted the flow of electrons to the plate. If the negative voltage on the grid was reduced, a small plate current did flow. Thus, a large plate current could be made to vary in direct proportion to a feeble grid voltage. The audion, de Forest's name for this modified Fleming valve, made it possible for a weak signal to be amplified (Fig. 3.5).

Actually, de Forest did not understand that the audion was amplifying; he thought it was simply an unusually efficient detector. Not until 1914 would the audion's actions be fully clarified. This work was carried out by Edwin Armstrong, an electrical engineer who would become a leading innovator in radio. His regenerative circuit achieved remarkable levels of amplification and showed how the audion could also be used for transmitting. De Forest and Armstrong became bitter enemies, spending much time in court contesting each others' patents.

Before the audion and its progeny could beget modern electronics, the three-element bulb would have to be perfected. De Forest's audions were made by McCandless, and each was unique. Some performed very well (and were sold at a premium), while others were just adequate or worse. What's more, they lasted for less than a hundred hours. At this time, of course, ordinary light bulbs also had pretty short lives. Clearly, some serious research was required to overcome these problems, but de Forest and his companies were unable to make the huge investments. With finances at a low ebb, de Forest sold most of his audion patent rights to AT&T in 1912. And so, improvement of the audion was undertaken during the next several years by the laboratories of industrial giants AT&T and General Electric.

AT&T's interest in the audion stemmed naturally from its attempt to expand long-distance telephone communication. With audion amplifiers ("repeaters") along the lines, phone signals weakened by their long travels could be boosted back to audibility. General Electric by this time had contributed other devices to the wireless art and was a big manufacturer of light bulbs. Thus, perfecting the light bulb and the audion could go hand-in-hand at GE, with the latter leading to new product lines.

The concerted efforts at GE and AT&T laboratories yielded impressive results. Within a few years, the audion's operating characteristics were improved dramatically, and its life was extended to thousands of hours. Audion amplifiers made it possible for AT&T to establish the first transcontinental phone link at the San Francisco World's Fair in 1915. De Forest also had a booth at the World's Fair. When he found out that AT&T's brochure describing the transcontinental triumph omitted mention of the audion, he became furious, vowing to "show them." He quickly produced his own pamphlet for fair-goers, a marvel of diplomacy for de Forest, that set the record straight.

Wireless and the audion were not the only technologies that went from laboratories and workshops into the marketplace during the first two decades of the twentieth century. During that period, the United States built an industrial infrastructure for the coming age of electricity.

3.5   A distinguished Lee de Forest, in later years, fondles an audion

Not only were staggering problems of power generation and distribution solved, but new administrative organizations—bureaucracies—grew to keep the large power systems functioning. Electricity, at first very expensive, was used in trolleys and streetcars, slowly making its way into factories and businesses. In 1910, only 10 percent of American homes had electrical service. It did not enter many other American homes until the late teens—and then only in cities. Even in the late twenties, few farms had electrical service.

After the telephone (which was battery-powered), the first important application of electricity in the home was, of course, for lighting. Improved light bulbs in the teens went hand-in-hand with the expanding power system.

Even before the turn of the century, techno-mancers had conjured up a host of new electrical products, and many were soon shown to be technically feasible. Beginning in the 1890s, small companies on the "entrepreneurial fringe" started marketing the first home electrical appliances, including heaters, flatirons, and assorted cooking devices. However, the ambitions of these companies, most of which failed, had outpaced the spread of electricity. An all-electric kitchen was publicized in 1904 (Fig. 3.6), boasting an oven, tea-kettle, chafing dish, waffle iron, and automatic coffee urn. By the teens, electric sewing machines, fans, heating pads, curling irons, and even cigar lighters had made an appearance, also somewhat in advance of adequate markets.

Sears Roebuck catalogs remind us that electrical devices of any kind were only to be seen in the homes of a few trendsetting consumers in cities, little more than toys for the wealthy. Though offering such modern conveniences as the Superba Ball Bearing Washer, The Princess Bust Developer, and countless pocket watches, Sears Roebuck did not yet sell a single electrical appliance—not even a lamp—as late as 1909. Even by the end of the teens, home electrical appliances were still pretty much oddities. But optimism in the industry sprang eternal, and soon it would be an optimism well justified.

3.6   An "Electric Kitchen Outfit," 1904

The financial innovations needed to put the new industrial products into the hands of consumers were already in place. Singer had been selling sewing machines on installment plans for decades. Soon most other sizable product purchases could be made on time.

Between the turn of the century and the Roaring Twenties, the revolutionary transportation technologies of the twentieth century also existed only in embryonic form. But progress was rapid. Bicycling was a craze and a mode of transportation in the 1890s, when the first practical automobiles were timidly setting forth from inventor's shacks. Surprisingly, the "horseless carriage" was touted then as a cure to the biggest culprit in urban pollution: horse droppings. Of the hundreds of tinkerers and engineers who established automobile companies in the early twentieth century, one name stands out: Henry Ford. In 1908 he marketed the first Model T (at $850); when the last one rolled off the assembly line in 1926, fifteen million had been sold.

The automobile and kindred vehicles like the Fordson tractor wrought many changes, mostly unanticipated. Among the least heralded but most significant was the vastly accelerated flight from the farm. The tractor was a curiosity in the first decade of the century, but by the late twenties a million tilled the land. The increased productivity created a huge labor surplus in farm areas, which fueled an unprecedented migration of Americans to the cities. This labor was absorbed almost effortlessly by the new industrial machine.

Ironically, Henry Ford and other entrepreneurs found through mass production a way at least partially to ameliorate the ills of industrialization. Their solution involved no social tinkering, as recommended by the utopian novelists, but would flow automatically from more efficient production. If products could be made cheaply enough, then even factory workers could afford them. Obviously, workers who shared in the wealth that industrialization was creating would be less likely to want revolution. Through assembly lines and relentless standardization (only the Model T was built), Ford was able to churn out cars that could be bought by almost anyone. Ford paid his workers $5 for an eight-hour day, which was regarded by other industrialists as "utopian." With that wage, they could indeed save up for a Model T ($400 in 1914). As other manufacturers in the teens and twenties adopted the "modern factory system" pioneered by Ford, the Consumer Age was born (see Chapters 5 and 7).

The rapid growth of the automobile industry was built upon the earlier foundation of iron and steel technology. These metals also made it possible for structures to soar above the city as never before. The first modern skyscrapers were built in the late nineteenth century in Chicago, and by 1908, with the construction of the Singer Building in New York, had reached a monumental forty-one stories. A distinctively urban skyline was taking shape.

Other inventions also got off the ground at that time. Human flight had been a wish since ancient Greece with the tale of Icarus, who ventured too close to the sun with wings of wax. During the latter half of the nineteenth century, powered flight had become a cultural imperative with an appreciable constituency, and feasibility loomed immediately on the horizon. In the last decades of the century, the pace of experiments—mostly supported by wealthy hobbyists or eccentrics—accelerated. The Wright Brothers demonstrated controlled, powered flight in 1903 and a few years later began building planes for the American military. The first two-way radio communication between airplane and ground took place in 1910 in New York State; later in the decade, during the First World War, airplanes would universally maintain contact by radio. Also in 1910, radio was installed in an English submarine and in Zeppelins.

In Germany, Zeppelins—distant relatives of the familiar Goodyear blimps—first flew in 1900; between 1910 and 1914 five of these airships carried 35,000 well-heeled passengers between various German cities. In the United States, the first scheduled airline flew in 1914 between Tampa and St. Petersburg, Florida, a distance of twenty-two miles, but it was many decades before air travel would have a place in the life of ordinary Americans.

While the new modes of transportation were breaking down barriers of space and time, physicists were quietly contemplating the universe, annihilating old ways of thinking about space and time.

In 1905 an obscure Swiss patent clerk published three papers in a German physics journal. One on the photoelectric effect was to earn a Nobel prize in physics for Albert Einstein. Another paper produced the Special Theory of Relativity and caused a theoretical revolution in physics and made its young author world famous. Special relativity proposed new and startling relationships between light, matter, and motion, predicting, for example, that as objects accelerate to near the speed of light, they become longer and do not age as fast. Perhaps because of these bizarre claims, Einstein's theory, a "free creation of the human mind," captured the public's imagination. Einstein became a celebrity scientist. One feature of special relativity had special import for radio. Einstein claimed that the propagation of light, and thus any electromagnetic wave, required no medium. Reluctantly, then, Einstein did away with the ether. Perhaps because aspects of the special theory were not verified experimentally until decades later, the term "ether" continued in use well into the 1930s. "Ether" retains a certain charm today, and so I employ it in discussions of early radio as it would have been used at the time.

By Einstein's time, a more mobile America was beginning to develop industrial products having greater portability. Items that could be moved from installation to installation or used by craftsmen in a variety of places were made in more "portable" forms, and they were advertised as such. By the turn of the century one could buy, for example, portable steam engines, portable boring and milling machines, portable search light equipment, portable electric drills, and portable forges. The new portable products, often on wheels, were more compact and lighter than their nonportable cousins. However, portable equipment was often more expensive and did not always perform up to expectations; portability had its price.

In warfare, portable communication devices were a long-standing cultural imperative. Problems of maintaining contact in the field were legendary and consequential. Some military historians believe that the Civil War would have ended in six months with much less bloodshed had the location of opposing troops been known in a timely manner. It should come as no surprise, then, that even before the turn of the century, wireless would be drafted by the military. Marconi himself had the opportunity to explore its use in the Boer War. He dispatched assistants and apparatus to South Africa with the British fleet, and radio was tested both at sea and on land. The results were mixed, owing mainly to antenna problems. Marconi solved the antenna problem in 1901 by building a wireless system into a Thornycroft steam-powered truck (Fig. 3.7). Atop the truck rose a huge metal cylinder, which served as the antenna. (It could be laid down on the roof when not in use.) Even when traveling at its top speed of 14 mph, this ungainly mobile radio still functioned, sending and receiving messages up to a distance of about twenty miles.

Radio also saw service on other fronts in the first years of the new century. Its eventual value was seldom in doubt, but the technology for practical portable wireless was not yet in hand. During these years in the United States, the military provided a sizable market for wireless apparatus of every sort and helped to sustain the efforts of the independent inventors, such as de Forest (Fig. 3.8).

The U.S. Signal Corps, a unit of the army, was very active in radio development. By the teens, the Signal Corps was already employing portable sets that could be carried on mules and set up quickly (Fig. 3.9, above). Other wireless stations were built into carts (Fig. 3.9, below) or special trucks; the Marconi company even sold a back-pack radio set. The manufacture of portable wireless apparatus for military use was so well along that the U.S. Patent Office employed the special category "portable sets" to cover new developments. In point of fact, most of the military sets, especially those made by Marconi, were technologically old-fashioned, consisting of cumbersome spark transmitters and receivers without audions.

3.7 Marconi's wireless-equipped truck, 1901

3.8 A de Forest portable outfit, 1909

3.9 Military portables. *Above,* display of
"Pack Sets" (early teens); *below,* Marconi
"Cart Set," 1917

When the United States entered World War I in 1917, the ingredients of more advanced wireless—long-lived transmitting and receiving audions that could be powered by batteries—were in hand. However, patent rights to the most important technologies were dispersed among a number of companies, which would not grant each other licenses. Nobody—not AT&T, not Marconi, not GE, not de Forest—could legally sell state-of-the-art radios. The U.S. government did find a way around this roadblock: it ordered the companies to cooperate and indemnified them against later claims.

During the war, GE, Westinghouse, and other manufacturers cranked out tens of thousands of military portables using state-of-the-art parts and circuits. In supplying tubes in unprecedented numbers for these sets, tube manufacturers, especially GE, made great strides in mass production techniques. The portable sets of World War I, used in the trenches, were the first U.S. radios manufactured on a truly large scale.

When the armistice ending the war was signed on November 11, 1918, in the forest of Campiègne, radio instantaneously signaled the good news throughout Europe, to ships at sea, and across the Atlantic. Wire services relayed the details of the German surrender to the newspapers, and by the next day virtually all Americans had learned that the war to "make the world safe for democracy" had ended with the Allies victorious. The brutal conflict had produced heroes of Andrew Jackson's stature, but none had to achieve his fame fighting in belated battles.

After the war, American radio entered a brief period of crisis. The temporary arrangement that enabled manufacturers to exploit each other's inventions without patent licenses would soon end. It would no longer be possible for any company to sell up-to-date vacuum-tube equipment. And there was also the Marconi problem.

American Marconi was a subsidiary of the British firm, and it held a virtual monopoly on commercial radio in the United States. Guaranteed a large market for its equipment by the Radio Acts of 1910 and 1912, which mandated the installation of radios on ships, American Marconi began to prosper. Just before the war, American Marconi had ordered from GE an enormous Alexanderson alternator, a generator that produced radio waves. American Marconi was temporarily nationalized by the navy during the war, and the alternator—then the world's most powerful—went undelivered. GE notified navy top brass and the president that this important high-tech item was about to fall into foreign (British) hands. The navy called together the principals and devised a plan of action that elegantly solved both problems. American radio would become an American monopoly. Following this plan, the major radio patent holders, including General Electric, AT&T, and United Fruit Company, pooled their patents and formed a new company in 1919: the Radio Corporation of America (RCA). (The founding companies were joined in 1921 by Westinghouse.) RCA was intended to operate the stations formerly run by American Marconi, which it bought out, and to sell radio apparatus built by GE and Westinghouse. At this time the markets envisioned for radio products were exclusively commercial, industrial, and military. The development of entertainment broadcasting and the exploitation of the vast market for home radio products that it opened up were not part of RCA's original mission.

Along with Marconi's thriving radio installations, RCA acquired a young employee, David Sarnoff, who would make RCA a major force as radio found new uses in the twenties and thirties. Sarnoff would also cultivate his reputation as an electronic visionary, a gifted techno-mancer.

In 1915, as a young but ambitious operator for American Marconi (Fig. 3.10), Sarnoff penned a now-famous memo to the company's vice-president and general manager. His proposal combined elements of Tesla's broadcasting system and Bellamy's music room into a unique vision: "The radio music box." In his own words,

> I have in mind a plan of development which would make radio a "household utility" in the same sense as the piano or phonograph. The idea is to bring music into the house by wireless. . . . The receiver can be designed in the form of a simple "Radio Music Box" [which] can be supplied with amplifying tubes and a loudspeaking telephone. . . . The box can be placed on a table in the parlor or living room, the switch set accordingly, and the transmitted music received.

Marconi, however, was not interested.

Though people in commercial radio thought Sarnoff's vision of entertainment broadcasting was preposterous, another group of Americans—amateurs and independent inventors—had already begun to realize this cultural imperative (Chapter 4). Very likely Sarnoff was familiar with these happenings and had simply transferred their vision of entertainment broadcasting to professional radio executives, who were totally unreceptive. Later, in 1920, the professionals would see dollar signs in the activities of the amateurs, and commercial broadcasting would be born. But the origin of radio as a medium of home entertainment lies firmly in the amateur camp. It is of interest here, then, that this same assemblage of undisciplined enthusiasts also built portable radios.

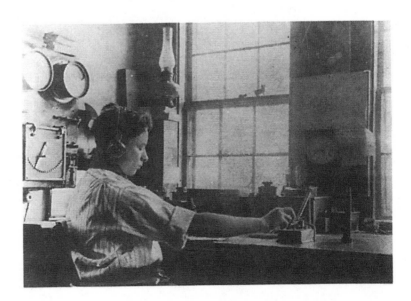

3.10 Young David Sarnoff at the key in a Marconi wireless station

4

Custodians of the Cultural Imperatives

HAD A WIRELESS-EQUIPPED SHIP PASSED CLOSELY BY Cobb Island, Virginia, in December of 1900, the radio operator might have plucked from the ether a surprising sound: Reginald Fessenden's voice. In actuality, the mile-long voice transmission had no eavesdroppers, and the sound quality from the spark transmitter was abysmal.

Reginald Fessenden was a talented engineer and inventor. Though he had once worked in Edison's invention factory, Fessenden was a free spirit who longed to realize his own visions. By the turn of the new century he had sharply defined one of these visions, a radio-*telephone* system. Fessenden wanted to offer point-to-point voice communication on a commercial basis.

As a result of his early experiments, Fessenden had concluded that high-quality transmission of the human voice was impossible with spark transmitters. A new source of invisible waves was needed; perfectly formed, continuous waves. Fessenden sought the solution in Tesla's alternating-current generators. He commissioned General Electric to build an alternator that could create a continuous radio-frequency wave, and he collaborated with one of GE's top engineers, Ernst Alexanderson, on this daunting technical challenge. In the meantime, Fessenden founded a wireless company to sell equipment and to furnish trans-Atlantic radiotelephone service.

The effort to build radio-frequency alternators began to pay off. Using an Alexanderson alternator, Fessenden established a two-way radio link between Brant Rock, Massachusetts, and Machrihanish, Scotland, in 1906 (Fig. 4.1). In the fall, a new 100 KHZ alternator of advanced design was delivered to Brant Rock. Unfortunately, the Machrihanish tower crashed in a storm a few months later.

Though nearly out of business, Fessenden found a novel way to use the new alternator. A few days before Christmas, he notified shipping companies that had installed Fessenden wireless apparatus to listen closely on Christmas Eve. Those who tuned in heard a special holiday program consisting of Fessenden playing the violin, singing, making a speech, and reading from the Bible. Radio operators, on land (as far away as Virginia)

4.1 Interior of Fessenden's Brant Rock Station, ca. 1905 (Alexanderson alternator is at lower right)

and at sea, were startled. Fessenden's Christmas Eve program of 1906 is widely regarded as America's first radio broadcast.

In one of those curious twists so common in the history of technology, Tesla, the man who had envisioned entertainment broadcasting, was unable to demonstrate its feasibility; on the other hand, Fessenden, whose interests lay entirely in commercial point-to-point communication, stimulated development of the first practical broadcasting technology (which exploited one of Tesla's earliest inventions). In a further irony, by the time entertainment broadcasting took hold in 1920, the Alexanderson alternator was on the brink of obsolescence, driven there by vacuum tubes used for generating continuous waves. Nonetheless, these whirring giants served radio well; after all, it was an Alexanderson alternator that General Electric had used as a lever to create an American radio monopoly through RCA.

Fessenden's achievement was a virtuoso performance in wireless technology (and maybe violin), but it was not the birth of broadcasting. The event generated no newspaper coverage, for Fessenden was indifferent to technological display and publicity, unwilling to humor or cultivate the masses. Even if his broadcast had been widely reported, it is doubtful that Fessenden's company or the other struggling American radio firms, such as United Wireless Telegraph Company, Fessenden Wireless Service, and American DeForest Wireless, could have succeeded in deriving any income from broadcasting. After all, in 1906, wireless receivers were confined almost entirely to ships and other special installations. The Panic of 1907 and the unfulfilled promises of the early promoters (radio telegraphy was not yet cheaper than wired telegraphy) dampened enthusiasm for new wireless ventures; as a consequence, many of the independent radio companies—almost all poorly managed—went out of business by the end of the decade. American Marconi survived and was poised to become dominant.

Although commercial radio was not yet the financial bonanza many had predicted, wireless was finding in amateurs a new constituency of dedicated enthusiasts who would realize the cultural imperative of entertainment broadcasting. America's seeming genius for tinkering constantly found new expression as boys and young men explored new technologies. Middle and upper class youth especially were drawn by popular literature into wireless because it was glamorous and promised to be a fun pastime.

As wireless became indispensable to commercial shipping, both newspapers and magazines carried accounts of radio operators—nearly all young men—whose vigilance at headphone and telegraph key had saved lives. When the ocean liner *Republic* was rammed in 1909 by the *Florida* in bad weather off Nantucket, *Republic* radio operator Jack Binns was able to reach the *Baltic*, which effected the rescue of more than 1200 people. Several days of front-page coverage elevated wireless (and Jack Binns) to the status of national heroes. Radio operators had become role models for American youth.

Juvenile literature fashioned new inventor-heroes for the new century; in these novels technical skill played a larger role than luck in the hero's success. A not insignificant number of books involved wireless. The first was John Trowbridge's *Story of a Wireless Telegraph Boy*, published in 1908. Soon popular boys' fiction series, such as the Motor Boys, the Rover Boys, and Tom Swift, included escapades with wireless. Later, the Radio Boys series showcased radio-based adventures. Even the *Boy Scout Manual* furnished information about wireless. Above all, wireless promised something extraordinary for ordinary people: the ability to communicate at a distance without corporate intermediaries.

The amateurs were, in fact, a breed apart—the early nerds. Most read voraciously about electricity and electronics, and built their own transmitters and receivers (Fig. 4.2). Using this equipment, amateurs spent countless hours scanning the ether, seeking distant compatriots. The fascination of this activity was elegantly captured by Susan J. Douglas in *Inventing American Broadcasting*:

Through wireless, the experimenter went through the looking glass, to a never-never land in which he heard the disembodied "voices" of ships' captains, newspaper men, famous inventors, or lovers passing in the

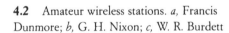

4.2 Amateur wireless stations. *a,* Francis Dunmore; *b,* G. H. Nixon; *c,* W. R. Burdett

night. This was an invisible, mysterious realm, somewhere above and beyond everyday life, where the rules for behavior couldn't be enforced—in fact, were not yet even established. The boy who entered it could, without detection, eavesdrop on the conversations of others. He could participate in contests of strength, power, and territory, and win them without any risk or physical danger.

Scarcely a decade before, the thought of children eavesdropping on "wireless" conversations would have seemed more like fantasy than science.

Though science and science fiction had long flirted, they wed in the works of Hugo Gernsback, unabashed promoter of radio as well as of futuristic visions. Gernsback was born in Luxembourg in 1884 and received a technical education. He emigrated to the United States in 1904, hoping to sell the first of his many electrical inventions. (He would eventually be awarded more than three dozen U.S. patents.)

In 1908 Gernsback founded the country's first radio magazine, *Modern Electrics,* in which two years later he published his own science fiction tale, *Ralph 124c . . . 41+,* set in the year 2660. Though regarded by modern writer Sam J. Lundwall as an "unendurable bore of a novel,"

Ralph contained some remarkable technological prophecies, including radar, tape recorders, and vending machines offering hot and cold food. (About the same time Gernsback prophesied pocket pagers and pocket telephones.) Gernsback also established *Radio News* (it began in 1919 as *Radio Amateur News*), which became the most popular radio magazine of the twenties. In addition, he sired the first magazine dedicated exclusively to science fiction, *Amazing Stories*, in 1926, and six other science fiction magazines as well. In 1929 this peripatetic publisher coined the term "science fiction."

For Gernsback, both his science fiction publications and his technical publications were part of one grand enterprise: science education for the masses—especially young men. According to biographer Mark Siegel, Gernsback believed that "the romance of fiction often conveyed the emotional impact of science better than did a purely technical article, and plot and characters often made science more understandable and memorable to young people."

All of Gernsback's works assumed that science (and radio technology in particular) had unlimited potential to benefit society: in radio magazines Gernsback popularized science; in science fiction magazines he popularized visions of new societies based on that science. Technology was to bring utopia. Hugo Gernsback appreciated, perhaps better than anyone, that science fiction—read mostly by adolescent boys and young men—could serve to create a cultural milieu favorably disposed toward scientific research and technological development. More than that, along with other media of popular culture, including movies, science fiction would become a means by which the visions of technically trained people, like Gernsback, could reach and influence a large constituency. Gernsback evidently understood that technological development both feeds into and feeds on popular culture.

Gernsback also believed that science fiction furnished specific inspiration for inventors, as the masthead of *Amazing Stories* unabashedly proclaimed, "Extravagant Fiction Today . . . Cold Fact Tomorrow." Science fiction contributed not only to literature but also to "progress"—which meant new technology. Thus, through his science fiction and radio magazines, he both stimulated and charted the growth of radio. He was also a techno-mancer par excellence; impatient with the current state-of-the-art, he was always seeing better things ahead. Through his many magazines, Gernsback was able to create constituencies for some of his visions, and they became cultural imperatives.

Boys attracted to wireless found this astonishing technology, which even promised to permit surreptitious communication with friends, surprisingly accessible. Beginning in 1901, *Scientific American* published plans for homemade wireless sets. The articles acknowledged that wireless *theory* was weighty, but a practical device could be built easily. The most expensive part was the spark coil ($6 in 1901). Those unable to afford a two-way set-up could, nonetheless, assemble a receiver for just a few dollars. And many did.

To help put wireless in the hands of amateurs and experimenters, Gernsback started up the Electro Importing Company in New York City in 1904. The following year the new firm was selling wireless parts by catalog, and it flourished. Later in 1905, Gernsback advertised a complete "wireless telegraph" (transmitter and receiver) in *Scientific American* (Fig. 4.3). Drawings of the "Telimco" wireless outfit generated thousands of orders, including several from New York department stores. In a telling commentary on the place of radio in American life at that time, the Telimco outfit was also carried by F.A.O. Schwarz, a large toy retailer.

A clever promoter, Gernsback was not averse to staging publicity stunts. In 1908, he installed his Telimco transmitter—with one-inch spark coil, spark balls, short wire antenna, American flag, and batteries—on the shoulders of a confederate who roamed the streets of downtown New York City with the transmitter sparking furiously. When the transmitter key was pressed, a bell on the receiver rang, as if by magic. In later years, Gernsback claimed that this demonstration set was "no doubt the first portable radio in the world."

Gernsback was among the earliest to appreciate that the construction of portable radios, his "walking wireless station," presented special design problems. When these problems were solved, the resultant equipment could be trotted out as a demonstration of technological progress. In the case of the Telimco wireless, the use of high-frequency waves dramatically reduced the length of the antennas (they were easily carried) and did away with the need for a ground, but reception was limited—from about 300 to 500 feet. For decades to come, others would use the portable radio to advertise technological virtuosity. And soon it would acquire other uses and meanings.

The portable radio was more than a novelty or a symbol to Gernsback; it was a device that someday everyone would own. That ever-smaller radios, capable of fitting into a coat and eventually shirt pocket, were in some sense inevitable was not a totally oddball idea, even in 1910, if one considers trends in comparable consumer products.

By the early twentieth century, Americans were probably the most mobile people on earth. A vast passenger train system interconnected virtually every hamlet and city. Commuter trains and trolleys—and later

4.3 Hugo Gernsback's Telimco wireless, 1905

the automobile—freed workers from the vicinity of the workplace during off-hours, and so bedroom communities and suburbs began to spring up. In addition, wealthy Americans were becoming accustomed to traveling in their leisure time, to lodges in the country, to the wild West, even to Europe. To accompany mobile people, inventors came up with "pocket" stoves for camping, portable electro-medical batteries, pocket knives, pocket razors, pocket gun oilers, and pocket flashlights. Soon there would be portable sewing machines and typewriters. In addition, numerous goods were sold in boxes or cases with handles for ease of carrying, including tackle boxes, camping outfits, toiletries for men and women, and musical instruments.

A more telling example comes from photography, which began in the second and third decades of the nineteenth century. By the 1860s and 1870s, the technology had matured considerably, but cameras were bulky and—except for those that itinerant photographers carted by wagon from hamlet to hamlet—quite sedentary. George Eastman changed all that in 1888 by producing the first Kodak portable camera that used roll film. This easy-to-use camera fit into a coat pocket and launched snapshot photography. By the mid-1890s, one could buy inexpensive novelty cameras the size and shape of a pocket watch (Fig. 4.4). Though watch-cameras did take pictures, they were scarcely more than toys, their design too severely compromised by the demands of miniaturization.

4.4 The Photoret pocket camera, 1895

Whether trends in other technologies, such as cameras, actually influenced Gernsback to promote portability in radios is not known, but the vision of radios small enough to be used by mobile people had evidently taken shape among wireless radio enthusiasts. Moreover, it is apparent from articles published by Gernsback in *Modern Electrics* and appearing in other radio and science magazines that the civilian portable radio was gaining a modest constituency. Portable wireless, including pocket-size sets, was viewed by many amateurs as a desirable product that was (by the standards of the time) eminently feasible. It had clearly become a cultural imperative. According to *Modern Electrics,* "Pocket wireless apparatus are now the order of the day." This was in 1911.

However, wireless technology was not especially well suited to making radios that were completely portable by today's standards. Ironically, the portable "wireless" required wires for operation. Thus, as in military gear, a portable wireless was somewhat compact, easy to move, and often had a handle, but once in place usually had to be connected to antenna and ground. As we shall soon see, some amateurs devised clever ways to circumvent the wire problem.

Modern Electrics and *Wireless Age* (published by American Marconi) not only included pictures and descriptions of the "home-brew" portables, but they also offered cash prizes for clever wireless set designs. The prizes often went to portables. Typical of these was the Keiling portable (Fig. 4.5, below), which received an "honorable mention" from *Modern Electrics* in 1912. The set was a compact crystal receiver built into a wooden case—the latter sold by Gernsback's Electro-Importing Company. C. Raymond Miller won a first prize in 1909 for his portable receiver-transmitter, which was built into a "dress suitcase" (Fig. 4.5, above). Another sophisticated portable set, constructed by H. Behlen, also was awarded a first prize (in 1913). It, too, was a crystal set in a box, measuring 9″ long, 6″ deep, 8″ high, and even had room to store the headphones.

Beginning in 1909, de Forest was selling audions to amateurs; by 1911, an audion could be bought for $4 from the Electro Importing Company. Not surprisingly, amateurs were soon constructing portables with audions, even before Marconi or the military. A nifty example is the 1-tube set put together by Stephen Anderson, Jr., in 1913. Because of the heavy load of batteries, the receiver required a "strong" box.

Miniaturization of home-brew portables was possible, but the penalties were severe. Because miniature coils and other parts were not in commercial production, the amateur had to make them. More commonly, though, miniaturization was effected by eliminating parts. This wasn't as hard as it sounds, because the most rudimentary receiver needed nothing more than a detector and headphone (and, of course, the pesky antenna and ground). Thus, tuning coils and capacitors, and so forth, were shed in the quest for miniaturization. Also lost in the process, however, were the abilities to pick up distant stations and to tune the receiver.

Even so, the cultural imperative of a pocket set was kept alive by periodic reports of mini-receivers. Charles L. Hedwell described one such set in 1913, built into an ordinary wooden watch box. Before presenting details of the set in *Modern Electrics*, he noted that "the name 'pocket wireless' is generally applied to any set which can be handily carried about. Here is a receiving outfit, however, which can actually be slipped into the vest pocket."

An even more diminutive wireless receiver was constructed by Louis C. Aldrich. His 1911 instructions for building a pocket detector begin as follows: "Get an old dollar watch case and take all the works out, including the stem." The watch-case detector was said to be "very handy for a portable outfit." These sets of ultimate portability obviously played poorly and were regarded as little more than novelties that advertised the cleverness of their makers. But there was the hope that someday such small sets would perform better.

Amateurs also contrived ingenious—if not always practical—solutions to the wire problem. My favorite is H. M'Cabe's "Portable Wireless Outfit," which used an umbrella for the antenna and a brass-bottom shoe to make the ground connection. Some creative tinkerers used fishing poles and kites to raise the antenna; one built an antenna into his hat. Portable radios did assume some strange disguises; one even masqueraded as a cane—a forerunner of the novelty sets so common today (see Chapter 14).

The most unusual home-brew "portable" of the era was William Dettmer's bicycle wireless of 1910, which included both transmitter and receiver. The antenna was erected on a fishing pole attached to the rear wheel, and a spring-loaded contact held a wire against the ground while the bicycle was in motion. The telegraph key and crystal detector were mounted on the handlebar. This set was clearly in the Gernsback–de Forest tradition of outlandish display!

4.5 Home-brew portables of the wireless era. *Above,* C. Raymond Miller's suitcase set, 1909; *below,* Keiling portable, 1912.

a

c

d

e

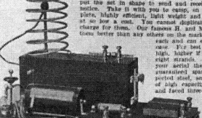

b

4.6 Commercial portables of the Wireless Era. *a,* Hunt and McCree 796 and 800, 1912; *b,* Hunt and McCree 801, 1913; *c,* F. B. Chambers & Co., 1913; *d,* Ellwood Manufacturing Co., 1913; *e,* Multi-Audi-Fone, ca. 1915

As might be expected, the interest shown by amateurs in radio apparatus of all kinds stimulated the commercial production of sets (but not of bicycle radios). They were manufactured, not by the wireless companies or electrical apparatus companies, but by tiny start-up firms, some run by amateurs. A few firms (like Zenith) would survive and become important in later decades. Ads for wireless apparatus at this time were confined mainly to radio and science magazines; the general public did not take part in the "crystal craze."

Seeing interest in sets that could be moved about with ease, a few companies actually began to advertise "portable" wireless. One of the earliest sets was sold by Hunt and McCree of New York, whose ads began in 1912. Their model 796, which could send and receive, was mounted on a solid oak base and came complete for $3.90 (Fig. 4.6a). Deluxe 2-way units were around $8.00 (Fig. 4.6a, b); a receiver only, "for the beginner," went for $1.75. Hunt and McCree outfits were acclaimed as "ideal sets for home, picnics, camping or boating." The F. B. Chambers Company also offered an unboxed "portable receiving set," mounted on a mahogany base (4-1/2″ × 12″), for $5 (Fig. 4.6c). Even in 1912 and 1913, though, these designs were not state-of-the-art portables, since they lacked carrying cases. Somewhat closer to the portable ideal (though not claimed as such) was "Ellwood's Wireless Receiving Set" (Fig. 4.6d). Advertised in 1913 as "the smallest set made," its mahogany case measured 6″ × 7″ × 6″. "Just the thing for your boy," blared the ad, though probably not at $30. One tiny commercial set was even promoted as a "pocket wireless" (Fig. 4.6e).

Because a few sets dubbed portables were made commercially for a home market in the early teens, this time can be regarded as the first portable age. However, it is equally clear that the manufacture of "portable" wireless was not an especially lucrative activity. The few set makers were probably encouraged by the spate of home-brew portables paraded in the pages of the radio magazines. Yet, the radio magazines, like all news media, featured "newsworthy"—that is, unusual—apparatus. The average amateur was doubtless content to explore the ether from his (or her) own room (Fig. 4.7). Clearly, the market for civilian portable radios, though enthusiastic, was rather small, too small to get serious attention from large, long established manufacturing firms.

Ads for commercial "portable" receivers, though uncommon to begin with, actually declined after about 1913. Most likely, amateurs who were really enthusiastic about portable apparatus built their own or bought a high-quality commercial set that, while not advertised as a portable, could be used as such. Since the better performing portables and ordinary receivers both required antenna and ground connections, there was little difference between them. Indeed, some of the nonportable sets, built into wooden cases, were eminently movable. And so, in the later teens, portables targeted at amateurs nearly disappeared as a distinct genre of home radio. However, improved portables were still being built by amateurs (and still winning prizes) and sets of all sorts were being *used* as portables in recreational activities outdoors. The Boy Scouts, in particular, took to wireless quickly, bringing along "portable" apparatus on their outings. Similarly, radio clubs were established throughout the country and often went on excursions with portable wireless.

The growth of amateur radio in the teens was nothing short of phenomenal. In 1909, Gernsback established one of the nation's first radio clubs, which a year later boasted a membership of more than 10,000. In 1914, the American Radio Relay League knit the amateurs into a nationwide organization. By that time, though, their freedom to transmit had been curtailed somewhat by the Radio Act of 1912. The act relegated their transmissions to higher frequencies (short waves), which the authorities thought unpromising for long-distance communication. In countless experiments, however, the amateurs proved the experts dead wrong. Indeed, much broadcasting and communication today take place on those very same frequencies charted out in the teens by the amateurs. The Radio Act also required amateurs who transmit to obtain a license. In 1917, there were more than 13,000 licensed amateurs. When the war came, amateurs were ready. Many joined the Signal Corps or navy to ply the airwaves for Uncle Sam. Gernsback's children were growing up. Later, when broadcasting boomed in the early twenties, many of these same people, their radio skills honed in the military, would enter commercial radio.

Amateur radio was—and continues to be—a "wholesome hobby" (Gernsback's term) for literate tinkerers, but in the early decades of the twentieth century, other pastimes had much more popular appeal. Americans, more than ever concentrated in cities, were beginning to spend appreciable amounts of time in "leisure" activities. Although the industrial workweek did decline from sixty-six hours in 1850 to about fifty-five by 1920, the new leisure was more the result of the urban-industrial lifestyle than of new-found time.

4.7   An amateur, Mrs. Chambers, at home with her wireless station, 1910

In cities large and small amusement parks sprang up; there were about 1,500 in 1919, at their peak. Amusement parks (and Vaudeville) declined in importance as moving pictures took hold. The first "movies" appeared in the 1890s in penny arcades, billiard halls, and so forth. By 1905, as projectors were perfected, special theaters were being built to show movies. More than 10,000 "movie houses" were serving 25 million people a week in 1910, as Hollywood began cranking out full-length feature films. Clearly, the movies were entertainment for the masses, patronized heavily by the working and middle classes. The wealthy, of course, continued to enjoy traditional forms of leisure, including concerts, the theater, opera, country club and college sports, and reading. The middle and working classes also took part in organized sports, particularly baseball. For most people, sports still involved active participation.

Social reformers saw the florescence of leisure activities among the working and middle classes as a necessary antidote to humdrum jobs. Increasingly, leisure was regarded as a way for Americans to find a happiness and fulfillment denied them in the workplace. Throughout the twentieth century, leisure activities have come to play an ever-larger role in defining and expressing individual identity.

Traditional leisure activities in the home, such as reading and making

music, continued to be important. In this period, mass-circulation magazines, such as *Colliers* and *The Saturday Evening Post*, were gaining a foothold. The ads in these magazines hawked the new national brand goods churned out across the land in Ford-style factories. A new culture, based on mass consumption, was beginning to take shape.

The phonograph still squawked better than it sang, but nonetheless attracted a loyal following. By the teens, phonographs (which now played discs) had entered many middle-class homes.

Styles of music were also changing, as the rural-urban migration brought together formerly distinct musical traditions. Prior to World War I, popular music—mostly sentimental ballads—was dominated by New York's Tin Pan Alley. Late in the teens the white songwriters and publishers of Tin Pan Alley began to popularize ragtime music, introduced by southern blacks who had moved into northern cities. This rhythmic music was a versatile medium for dancing, and so the nation was swept by a dance craze (Fox trot, turkey trot, etc.). The resultant dance halls became quite popular among youth. Stimulated by the dance craze, phonograph sales reached two million in 1919 and remained strong in the twenties (Fig. 4.8). Significantly, popular dance and popular music were wed.

Columbia
Grafonola

The joy, the intense, wholesome joy which a Columbia Grafonola brings into your home will make all your family more keenly alive to the spirit of Christmas.   It is a double joy the Grafonola gives.   There is the joy of immediate possession and the joy of anticipating the ever-new pleasure of good music for year after year to come.

Begin early to shop for your Christmas Grafonola.   Columbia dealers specialize in making Grafonola buying a pleasant, holiday sort of business for you.   You will be as welcome in the Columbia store as your neighbors who come to pay you a Christmas morning call in your home.

Columbia Grafonolas are priced at $18.00 to $250.00
Period Designs up to $2100.00

COLUMBIA GRAPHOPHONE COMPANY, New York

4.8   Ad for Columbia Grafonolas, 1918

Both music and sport—enjoyed vicariously—would in the decades ahead achieve an even greater importance in the lives of ordinary Americans as a result of radio. Eventually, listening to the radio became nearly everyone's hobby, but first there had to be entertainment broadcasting.

When de Forest immodestly entitled his autobiography *The Father of Radio*, he perhaps had the audion less in mind than his pioneering broadcasts. According to de Forest, in February of 1907 he began experimental broadcasts of music and voice from his laboratory and from the Telharmonium Building in New York. Would-be listeners scrambled to buy apparatus, and de Forest obliged by opening a radio supply store in 1909.

De Forest picked up radiotelephony where Fessenden left off, but he had a much different vision of the new medium's potential. Though aspiring to be (and he sometimes was) wealthy, de Forest also wished to democratize the leisure activities of the upper classes. Like Bellamy, he envisioned a device that could deliver into every American home the finest music and opera. De Forest believed fervently that the new technology of radiotelephony could bring this vision closer to reality. Between court appearances (for patent infringement and stock fraud), he diligently pursued his vision of entertainment broadcasting.

As early as 1909, de Forest had pondered how to make broadcasting pay. Among his suggestions were advertising and a monthly fee on receivers. (Curiously, the latter alternative was adopted by the British.) De Forest's broadcasts did not actually cover their direct costs; rather, they served his larger interests: to test new tubes and equipment, to demonstrate the popularity of broadcasting among listeners, and to stimulate sales of his radios and supplies to amateurs.

Between 1909 and 1919, de Forest set many precedents in entertainment broadcasting. His programs included news, regularly scheduled music (beginning in 1915), sports, and presidential election returns (1916). In 1910 he transmitted Enrico Caruso live from the Metropolitan Opera House in New York City. Perhaps his most interesting broadcast took place in 1909, when his mother-in-law, Harriet Blatch, spoke on behalf of woman's suffrage.

When not broadcasting, inventing new vacuum tubes, or getting into legal troubles, de Forest made some singular contributions to the portable art. Best known are his military sets, which contained audions, and were of advanced design. It is less well known that de Forest invented the "fountain pen radio," a plan for which was published in 1917 in *Electrical Experimenter* (another Gernsback publication). Actually, the one-audion receiver was much larger than a pen, but it had some fascinating features. At the tip was a built-in earphone; the listener merely stuck that end in his ear. The antenna wire ran up one sleeve and down the other, then entered a hollow cane containing a "spiral aerial." The ground wire, which also went into a sleeve, emerged through a trouser leg and attached to a metal plate on the bottom of a shoe. De Forest suggested that this design might be ideal for tracking down "spy radio stations" (it was, after all, wartime)—as if someone walking around with a cane and a large pen in his ear would not be conspicuous. Needless to say, the "fountain pen radio" did not enter commercial production.

After the war, de Forest continued to invent, but his broadcasting career was terminated when federal officials in 1919 shut down his transmitter, which he had moved illegally. Long before this setback, however, the germ of entertainment broadcasting had spread widely among amateurs. Throughout the land in the teens, amateurs and radio experimenters were sending voice and music into the ether.

For much of this period, radio apparatus was still too crude and finicky to generate mass appeal. However, within a few years after the war, enormous vacuum tubes were put into service that could transmit flawless continuous waves, and vacuum tube receivers achieved a surprising state of perfection. Radio hardware was at last ready for prime time. Entertainment broadcasting was about to expand beyond the absorbing, self-indulgent pastime of amateurs to reach millions of Americans seeking novelty in their leisure time.

The turning point came in 1920. Frank Conrad was a radio amateur who worked as an assistant chief engineer for Westinghouse in Pittsburgh. (Conrad's prior claim to fame was perfection of the watt-hour meter used to monitor electricity consumption.) For several hours each evening, he tinkered with his transmitter and played records over the air. As word spread of these entertaining interludes in the ether, more and more amateurs tuned in. In 1920 the Joseph Horne Department Store began selling rudimentary radios that, it was claimed, were perfect for picking up Frank Conrad's musical programs. They were crystal sets and cost $10 and up.

One day in September Harry P. Davis, a Westinghouse vice-president and Conrad's supervisor, happened upon an ad for the Conrad-ready radios. Westinghouse was not yet in the radio group that formed RCA and had been looking for new markets to redeploy the huge production capacity developed during the war. Here was a novel idea that Westinghouse might exploit: selling radios to the public. This chance discovery, then, catalyzed the process that was leading inexorably to a broadcasting boom and to the manufacture of home receivers on a large scale—by Westinghouse and hundreds of other companies. That same September, Westinghouse invited Conrad, who had been doing the radio work in his garage, to set up a station in a company building. This station was quickly licensed and christened with the call letters KDKA. It became this country's first commercial radio station dedicated to entertaining the public (and stimulating radio sales). KDKA's premiere broadcast, of the Harding-Cox election returns on November 2, 1920, was perhaps heard by a few thousand people.

As a result of nationwide newspaper coverage of KDKA (and the free publicity given Westinghouse) many corporate executives became convinced that radio could reach people by the millions. In no time at all, hundreds of companies—from the *Chicago Tribune* to AT&T—had joined Westinghouse in starting their own stations. By 1923, there were more than five hundred radio stations around the United States licensed for entertainment broadcasting.

At first, the companies that founded radio stations—newspapers, department stores, hotels, and so forth—regarded them as good public relations, a service that would repay the owner through goodwill. Soon it was learned that running a radio station was an expensive proposition; goodwill did not immediately pay the electric bill or hire first-rate talent or announcers. And so broadcasting's first years—well into the late twenties—witnessed many experiments with ways to make radio pay; only slowly did the modern way—advertising—take hold. During the radio craze of 1920 to 1926, most stations still did not accept advertising. As late as 1928, the National Association of Broadcasters forbade commercials in the evening hours.

As Susan Douglas has noted, the triumph of KDKA and the rush to the ether that followed represented a transfer in the control of the new medium—entertainment broadcasting—from amateurs and experimenters to commercial interests. Though entertainment broadcasting had been a cultural imperative for decades and was brought into being by the amateurs, corporate America established the institutional framework within which it would finally grow and prosper. The realization of this vision—shared by Bellamy the novelist, Tesla the engineer-magician, Sarnoff the radio man, de Forest the physicist-inventor, and countless amateurs—was imminent.

5

The Radio Craze

FOR MANY PEOPLE, MENTION OF THE ROARING TWEN-
ties—when prohibition prevailed—conjures up strong images of speak-
easies, gangsters in obscenely long cars, flappers and jazzmen, the
Charleston, the Teapot-Dome scandal, laissez-faire Republican presidents, stock-market mania, and so forth. The movies and dull history
tomes give us distinctive images of that decade; what they miss, however,
are fundamental changes in the patterns of everyday life.

The twenties roared, because in that space of ten years, the lives of
ordinary Americans were transformed more rapidly than in any previous
decade. Life was a giddy ride, with endless new wonders around the
corner. Soured on internationalism, Americans turned inward, building
on the new industrial base a consumer society of unprecedented scale.
Americans could not change the world, but they could change their lifestyles. T. M. Kando, writing in the seventies, has observed that "the roaring twenties were culturally in many ways similar to our own era today.
The hedonistic, glamorous, and make-believe world dangled before the
masses by the media . . . found its first massive manifestation [in that
decade]. Crass materialism, blatant sexuality, the cult of youth, the il-
lusory pursuit of excitement and success, these were the dominant values
of the twenties, as they are today."

For the first time in history, more than half the U.S. population lived in
towns of 2,500 or more, and urban values and activities began to dominate America. Automobiles had become ubiquitous, competing with
streetcars for space on narrow city streets. They were made by dozens of
manufacturers, most no longer in business, like Pierce-Arrow and Hup-
mobile, Reo and Rickenbacher, and Peerless and Packard. Two decades
earlier, autos had begun as playthings of the rich. By the twenties, under
the influence of Henry Ford, they were reaching the masses. Other consumer products, including radio, would also reach into every stratum of
American society.

The twenties represents the first transition to the modern age, in
which emerged a universal consumer consciousness—the insatiable ap-
petite for new things—shaped by mass-circulation magazines, news-

papers, and movies. In these media, large numbers of middle- and even working-class consumers could actually see, for the first time, exactly what kinds of clothing, cars, furniture, and appliances the wealthy possessed. They could also be enticed into believing that America's industrial might—and installment buying—would bring these products within their grasp. The bug of acquisitiveness that had always afflicted the well-to-do was now infecting everyone. Yet, the twenties, as a transitional period, was an uneasy mix of the old and the new.

At the beginning of the decade, home electrification was far from complete. In 1921, for example, only 44 percent of the residences in Chicago had electricity. As electrification reached the last city homes in the late twenties, a spate of new electrical things followed. Many had been kicking around in some form since the 1890s, and they now finally acquired the cachet of practicality. By the mid-twenties, vacuum cleaners, electric irons, electric ranges, electric coffee pots, and even electric washing machines were beginning to enter a few middle-class homes. Even staid Sears Roebuck carried in its catalog a breathtaking assortment of plug-in devices, including waffle irons and corn poppers, hair dryers and curling irons, egg cookers and massagers, paint sprayers and electric drills. In just a few years, even these "terrific" products would be judged, in comparison to their descendants, as crude and impractical.

Dedicated trendsetters could purchase still more marvelous goods pictured in color ads in *The Saturday Evening Post*. The health-conscious, for example, might buy the "Battle Creek Health Builder" (a large belt vibrator), while the "Eveready Sunshine Lamp" offered the vain a midwinter tan. Compulsive (and wealthy) housekeepers could have a Finnell Electric Floor Machine: "It waxes, it polishes, it scrubs." By the end of the decade, trendsetters bought electrical products never before seen in the stores, including televisions and electric refrigerators.

Radios (using technology of the late teens), which were about to burst into homes around America, differed in one important respect from the other new electrical things: they could not be plugged in because they worked on direct current (DC). Though Westinghouse had won the current wars with alternating current (AC), DC systems did not completely die out. DC was actually preferable in some applications because it could recharge batteries (all batteries are DC). Ships, for example, had DC power—as did automobiles. In fact, radio technology (the tubes in particular) had matured in applications—mobile and portable gear, for example—based on DC. That is why home radios of the early and middle twenties required batteries. In effect, the receivers of the radio craze were but barely reformed portable radios of the late teens. Only after radios started selling in sizable numbers, and the pains of battery-only operation in the home became evident, did companies begin developing the technology for making plug-in radios. In 1927, AC sets made a big splash in the market (see Chapter 7).

Technological development, especially in radio, was pushed hard by the rising consumer sector; however, the role of independent inventors like Fessenden and de Forest was somewhat diminished, as large corporations with well-financed research laboratories plunged into consumer products. By 1930, there were more than a thousand corporate laboratories—a tenfold increase in about a dozen years. The corporate laboratories excelled at perfecting a product and gearing it to mass production; they were less capable of making the big breakthroughs. That is why, even in the twenties (and to this day) a disproportionate number of significant inventions and products continued to be contributed by independent inventors as well as by small companies.

The national political environment of the twenties was conducive to the expansion of large corporations; two Republican presidents allowed big business to have its way with America—no questions asked. In 1923, when the "radio craze" was in full swing, Warren Harding's presidency had already come to a premature close; famous for the Teapot-Dome scandal, it was otherwise undistinguished. Harding was replaced in the White House by Calvin Coolidge, who was best known for his prodigious napping.

The electric refrigerator was still a few years away, so most families kept their food cold in an "ice box," whose charge was regularly replenished by the ice man. Along with the "church key" (for opening bottles and puncturing cans), the ice pick was an essential item of household technology. Ice boxes being small, perishable foods were bought daily from the corner grocer. However, dairy products were delivered to the home by the "milkman," often driving a horse-drawn truck. Clothes were washed in a tub or with a ringer washer and hung out to dry in the fresh air. There were no laundromats, but in big cities commercial laundries took in washing.

Though life was different in this era, there were some comforting continuities. Campbell offered 21 kinds of soup in a can surprisingly similar to today's. Americans concerned about getting enough fiber in their diet could buy Pillsbury's Health Bran and make "tempting golden-brown muffins, delicious cookies, and rich, wholesome bread" (recipes were on the package). American Express was already promoting paranoia to sell its Travelers Checques in this 1925 ad from *Travel* magazine: "Millions of dollars, carelessly carried in the pockets of the people, are thus lost annually in the United States. It is this sort of carelessness that makes crime easy—and *Crime*, and *Carelessness*, cost our people $3,500,000,000 last year." Listerine, "a pure, saturated solution of boric acid skillfully and always uniformly blended with healing, fragrant oils," was essentially a nineteenth century patent medicine looking—in 1923 and today—for legitimacy and new uses in a changing world. In addition to combating halitosis as a mouthwash and gargle, Listerine was claimed to prevent sore throats, to be an effective skin bracer and aftershave, an exhilarating scalp massage that "is effective in combating dandruff," and a "safe non-irritating deodorant." Though familiar, these threads of life from the 1920s were woven in a strikingly different tapestry. Perhaps nothing illustrates this better than radio itself.

Radio was the first *electronic* device to enter the American home. Often it arrived as component parts, with some assembly required. Components could be purchased at Woolworths or at other "five and dime" stores and through specialty catalogs (Fig. 5.1, above). Even Sears Roebuck had a catalog devoted to radio, and offered reduced prices on common

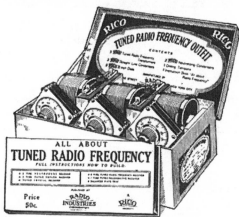

5.1   Many home radios of the twenties were bought as parts or kits. *Above*, cover of RASCO catalog, No. 11; *below*, Radio Industries Neutro-Reflect Kit

tubes—only $3.65 each. More than a dozen new magazines, including *Radio Age, Radio Broadcast,* and Gernsback's *Radio News* (the most popular, by far), supplied information about radio developments and circuit diagrams to amateurs and the new "radio enthusiasts"; they also carried ads for parts and radio kits (Fig. 5.1, below). By the middle of the decade, when at their peak, the radio magazines had a combined circulation of more than one million.

Newspapers and mass-circulation magazines carried articles about radio, and *Colliers* even contained a column, "Picked Out of the Air," by radio hero Jack Binns. Ads for radios were common in magazines that reached into every sector of society, from *The Literary Digest* to *The Farm Journal.* And so, in the early twenties, radio became the rage (Fig. 5.2).

The coming of the radio craze opened up vast opportunities for that most American of institutions, the entrepreneur. Radio shops appeared in every city, and some—like New York—had hundreds, stocked with countless new brands. Because even the most complex sets had only a few dozen components whose assembly required little equipment or talent, many people went into the radio manufacturing business. Regrettably, the combination of ingredients needed for success—a good product, business acumen, adequate capital, and effective advertising and distribution—were uncommon. As a result, hundreds of manufacturers came and went, the majority rather quickly. They left behind some memorable names (and a few memorable radios): DeWitt-LaFrance Company's "Superadio Reatodyne 6," David Grimes's "4DL Inverse Duplex Reflex," and Kodel's "Logodyne Unitrola," all 1925 models. None of these companies produced radios again. Of 153 reasonably successful companies (advertised nationally for at least two years) that made radios in 1925, 58 had ceased business or ceased selling radios by 1929. The decade of the Depression took many of the rest.

5.2   Radio is a sensation in the early twenties. Shoppers have the opportunity to listen to radio in New York's Bruck-Weiss Store, 1922

The one manufacturer most closely linked to the romance of 1920s radios is Atwater Kent. The man who gave his name to that company was already a successful industrialist, manufacturing electrical devices for cars, when he turned to radio in 1921. His earliest "breadboard" sets, which the purchaser assembled on a handsome slab of wood, today are prized collectibles, selling for more than $500. (The early Atwater Kent sets were sold as kits in order to circumvent RCA's lock on radio circuit patents.) Advertised extensively in national magazines (Fig. 5.3), the Atwater Kent line created a devoted following by stressing quality: "In Atwater Kent Radio," said a 1925 ad in *The Farm Journal*, "there is quality beyond question; there is scientific, precise design and master workmanship." Looking at these sets today, we see that the claims were not exaggerated. Atwater Kent did use components of the highest quality, beautifully designed and made in his own factories; every part seems to reflect an unswerving attention to detail.

Although Atwater Kent components were well made, the finished radios did not use state-of-the-art circuits—far from it. The real radio mavens cared about such things, for magazine ads constantly bombarded the reader, making promises about revolutionary circuits that to the uninitiated must have seemed mystifying. These circuits ranged from "neutrodyne" and "unidyne" to "superregeneration" and "duplex reflex." Only a few of the many "improved" circuits of the early twenties would have lasting importance. Among these is the superheterodyne, invented by Edwin Armstrong while he was with the U.S. Army Signal Corps in France during World War I. In a superheterodyne circuit, the incoming high-frequency radio signal is mixed with a lower frequency signal generated by the radio itself. The combined signal of constant intermediate frequency can then be passed through one or more stages of high-efficiency amplification. Although the principle of heterodyning had been thought of before Armstrong, only he perfected it to the point where its great superiority became obvious.

In 1922 David Sarnoff, already general manager of RCA, saw a demonstration of Armstrong's superheterodyne receiver and was very impressed. Initially, however, the RCA people were not receptive to buying a circuit from someone outside the radio group. So, Sarnoff arranged for Armstrong—whom he had known for a decade—to make a presentation to RCA and Westinghouse officials at the apartment of Owen D. Young, then chairman of the RCA board, early in 1923.

The radio Armstrong brought along to demonstrate the virtues of his superheterodyne circuit was a portable (Fig. 5.4). According to Armstrong, "The set measured 18 by 10 by 10 inches . . . and was completely self-contained—the batteries, loop antenna, and speaker mechanism being enclosed in the box." With this portable radio, Armstrong, who had a flair for the dramatic, soon clinched his case. According to George Douglas in *The Early Days of Radio Broadcasting*, "Armstrong stepped off the elevator carrying the set under his arm in full operation with an opera program in progress." Apparently, the sight (and sound) of a completely portable radio playing well without external antenna and ground was enough to convince RCA officials that the superheterodyne circuit had great merit. RCA worked out a deal with Armstrong and began pro-

5.3 Ad for Atwater Kent radios and speakers, 1924

ATWATER KENT

RADIO

You'll Never Forget the Night

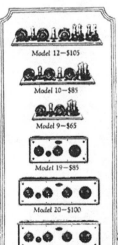

Model 12—$105

Model 10—$85

Model 9—$65

Model 19—$85

Model 20—$100

De Luxe Model—$120

YOU'LL never forget the night you first tune in your ATWATER KENT Radio. The thrill of it will live in your memory— the sheer delight of filling your room with living voices or the music from an orchestra perhaps a thousand miles away.

Its clear reception, and the ease with which you can bring in distant stations will be a revelation to you. An added pleasure will come with the knowledge that no one possesses better radio than yours.

ATWATER KENT craftsmen, guided by the experience of skilled engineers, have fashioned the finest

materials that money can buy into ATWATER KENT Radio. You will find it combines every feature that means radio satisfaction—unusual selectivity, sensitiveness, distance, volume and tonal quality.

The ATWATER KENT dealer near you will gladly help in the selection of your radio. There is a price, size and style for everyone.

Instructive literature on request

ATWATER KENT MANUFACTURING COMPANY
4713 Wissahickon Ave., Philadelphia, Pa.

Model R
$15

Model M
$28

Model L
$20

THE true worth of a loud speaker is judged by its faithful reproduction of broadcasts. In ATWATER KENT Loud Speakers each kind of material used, each detail in design is there for a purpose—to bring about a tone that is pure, clear and natural.

ATWATER KENT Loud Speakers bring out the best from any set.

Skilled engineers and master workmen have set a new standard in their production.

THINK OF WHAT IS BACK OF IT

ducing pricey superheterodyne receivers in 1924. After 1930, the super-heterodyne quickly became the standard of the radio industry, and remains to this day the basis of all radios and televisions.

Constant competition among new circuits, improved components, and incessant claims about breakthroughs just around the corner sent a subtle message to the majority of Americans: radio technology must still be fairly crude as attested by the rapid pace of new developments. As the technically sophisticated market became saturated, radio makers began to send other messages to the masses. "Why wait for the ultimate," a Tuska ad asked, "*it never comes!*" Don't wait for further progress, other manufacturers counseled, because good radio is here today. Not until the end of the decade was that claim finally true. Although many intrepid tinkerers built their own receivers, the preassembled sets sold well (Fig. 5.5). Even the seeker of a ready-made set had to become slightly immersed in the "radio art," at least learning enough of the new lingo to converse with salespeople. For example, a "peanut tuber" was not something eaten salted, with beer, but a one-tube receiver using a small (peanut) tube. Four-tubers and five-tubers, having precisely that number of tubes, were among the more expensive—and usually better performing—sets. A prospective customer had to understand "sensitivity," how well a set could bring in weak stations. (More tubes usually meant more stages of amplification and thus greater sensitivity.) "Selectivity" was also important, for it indicated a set's ability to discriminate stations having nearby frequencies.

5.4 Edwin Armstrong and his super-heterodynes. *Left,* inventor of the super-heterodyne circuit poses with an early demonstration set; *right,* the superheterodyne portable used by Armstrong to sell the circuit to RCA

5.5 Westinghouse Aeriola, Sr., a pre-assembled set with one tube, is shown in this 1922 publicity photo

Price was a prime consideration, and there was no lack of choice on that score. RCA ads bluntly claimed in 1922 that "there's a radiola for every purse," providing radios from $18 to $350; by 1927, the range was $82.75 to $895.00. For people of great means Zenith had a line of radios in "Deluxe Art Model Cabinets from $500 to $2,000."

To put these prices into perspective, it should be appreciated that, in 1925, a room for two at an inexpensive hotel in New York City cost $4; at an expensive hotel it was $12. A copy of *Collier's* weekly magazine cost a nickel. A dozen pulp novels, "every one a ripsnorter," went for $1.98. However, a copy of Gernsback's *Ralph 124c*—with a TV on the cover— went for $2.15 (Fig. 5.6). Oldsmobiles were $1,000 to $2,000, but some Fords and Chevrolets cost less than $600. Sears sold mail-order houses (some assembly required) for under $500, while a nice house in the city, already built, fetched from $6,000 to $12,000, depending on the neighborhood.

Two thousand dollars for a radio was a king's ransom; even $200 was steep for the average worker, amounting to several month's salary. Perhaps that is why Crosley sets ("Better—Cost Less"), the cheapies of the industry, were so popular; the Crosley "Pup" was a one-tuber that sold for $10. Inexpensive crystal sets were still available for as low as $1. Those who spent in excess of $120 on a radio (the price of Atwater-Kent's 1925 DeLuxe Model) were clearly buying more than an entrée to the ether. Freed-Eisemann was not subtle about this, claiming in ads (Fig. 5.7) to be "The radio of America's finest homes; always first in social prestige."

The would-be radio owner soon discovered that the price of the radio itself was scarcely a down payment for access to the ether. One-tubers and two-tubers, and so forth, were in reality no-tubers, because most companies sold their tubers tubeless. The outlay for tubes could be considerable, around $5.00 each in 1923 and 1924.

5.6   Hugo Gernsback's ad for *Ralph 124C . . . 41+* in *Radio News* (note the television receiver)

5.7   Freed-Eisemann ads had rather blunt snob appeal

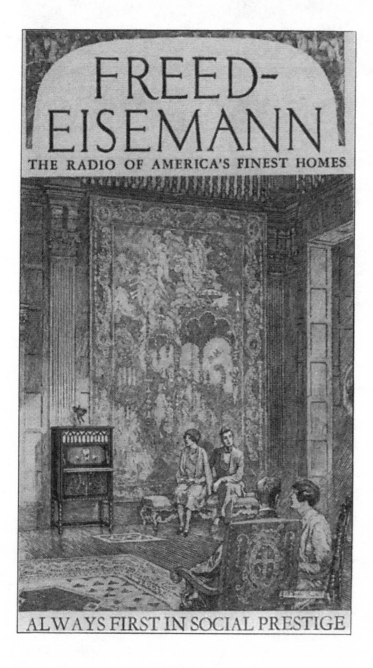

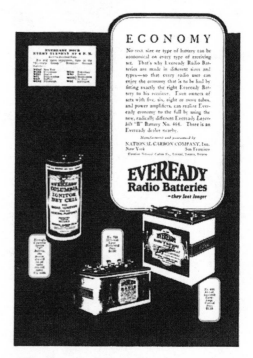

5.8   Eveready radio batteries, 1925

5.9 Crosley Musicone speaker shown outdoors in a publicity photo

To get the tubes to light up and carry current required batteries (called the A battery). Many early sets had tubes with inefficient, high-drain filaments (like the 201, which consumed 1.25 watts). To power these one bought a rechargeable storage battery, like those still used in automobiles, for from $15 to $30. Dry batteries supplied plate current (the so-called B battery) and added from at least $4 to $12 more to the outfit (Fig. 5.8); some radios also had a C battery (add $1.50). The last "accessory" needed for the radio system was the headset (or speaker), which was available in many varieties (Fig. 5.9). Headsets sold at $4 to $12.00, but speakers were somewhat higher, $10 to $50 for a good one. Add $5 or so for antenna, ground, and battery wires, and the preassembled radio was ready to be taken home and assembled.

Because secondary schools still taught the "manual arts" in those days (the proper use of tools and so forth), the prospect of putting together the radio was not really so daunting to the average enthusiast. However, there were pitfalls to snare those untutored in the ways of electricity. By connecting the higher voltage B battery where the A should go, the tubes burned out—instantly. Stringing the antenna, between house and tree for example, was a dangerous chore, especially if the ladder was rickety. Rigging up an indoor antenna could also be a challenge, as the following account from a Tulsa paper suggests:

> After his radio aerial had broken loose accidentally Tuesday, J. H. Bolinger tied a connecting wire about the foot of his dog and heard a concert being radiocast from WKY, an Oklahoma City station. . . . By hooking up the dog, his wife, daughter, and a friend, Bolinger said, he enjoyed a concert from a Los Angeles Station. Bolinger stated that, although the dog was a wire-haired fox terrier, the wire hair does not seem a necessity and that any dog will serve just as well as an aerial for short-distance receiving.

Eventually, the radio tinkerer connected all the wires in the right places, grounded the set, and turned it on. The first sound was apt to be a blast of static, a whistle, or announcers from several stations babbling at once. Tuning was an art like ballet that required stern concentration and was acquired only after much practice. In the early sets, several tuning capacitors had to be adjusted more or less at the same time. In practice, that meant turning each knob, one after another, a fraction of a degree and waiting for the result—usually more static or squeals, but sometimes a station.

A skilled operator was not enough to guarantee that voices and music would issue forth with clarity. Reception was dogged by two radio gremlins of that time: static and "birdies." Static is a term still familiar, and so is the sound, found readily between today's AM stations. The causes of static were poorly understood in the twenties, but there were some obvious culprits. The passage of an electric streetcar nearby, a low-flying plane overhead, or even a car moving by the house created a crescendo of static that could overwhelm even strong local stations. Static also came from electric motors on a then-rare vacuum cleaner or electric sewing machine. The most consistent cause of static, of course, was lightening. Birdies were the chirps and whistles picked up when a neighbor's set was radiating because of unshielded oscillator circuits. To get a feel for birdies, one can bring an AM radio close to a television; radiating circuits will cause birdies to abound.

In addition to the gremlins of static and birdies, assorted (but unnamed) hisses and whistles and squeals rounded out the auditory adventure. Radio obviously remained to be perfected, jibed journalists and cartoonists. The following commentary appeared in Jack Binns's radio column in *Collier's* of November, 1924: "Harold Speake of Chicago suggests that the stockyards have at last found a market for the hitherto only unsalable by-product of hogs. Some radio manufacturers are the customers, he alleges." For the most part, the gremlins and such were beyond the control of the "radio nut," who was advised to cultivate the virtue of patience.

Having become knowledgeable in the ways of radio, and having survived its gremlins, the enthusiast could look forward to more expenses and mishaps. Acid-filled storage batteries could be spilled on carpets (which were wool), with unpleasant results. A "Tufglas" battery tray was available for $1.50 to protect floor coverings from just such a catastrophe. Outdoor antennas were another source of misery, especially when they attracted lightning: if the house survived, the radio inevitably didn't.

When disasters did not strike, the desire to upgrade equipment often did. Wealthy "radio fiends" thought nothing of replacing their sets with a new model every year or two. Most, however, obtained improvements at slightly less expense. A bigger and better speaker probably came quickly, especially as improved cone varieties became available. The storage battery had to be regularly recharged, usually by a service station ($1.00 for pick-up, charging, and delivery); an obvious solution was to buy a charger ($18.50 and up) and do it at home for a nickel's worth of electricity (Fig. 5.10). The B battery could be replaced by a "B battery eliminator"—for $27.50 and up. Radios with few tubes gave feeble sound, inadequate for driving a speaker. An accessory amplifier that required additional battery power remedied that defect for about $20. An elegant solution was the Aero B Amplipower, which combined an amplifier and B power supply—for $65 (less tubes). The popularity of these add-ons is difficult to gauge, though Balkite claimed in 1926 that 10 percent of American radios were equipped with its chargers and eliminators.

Despite the problems with radio and its seemingly endless costs, receivers did sell. Although many estimates of radio sales are available for the twenties, they are all guesses. For mid-1925, these "guesstimates" of total U.S. radios range from three to eight million sets. These numbers testify both to radio's rapid rise and to its untapped potential, for only a small fraction of American homes—probably fewer than 10 percent—enjoyed radio at that time. Because radios, being battery-powered, could be used in any home (electrified or not), and because they were available at prices affordable by nearly every family, the penetration of radios at this time—though thin—was unusually broad. What seemed essential was that a family have one adventurous soul willing to learn the lingo and undertake the assembly and operation of the receiver. The many returning soldiers, of diverse social class (some of them former amateurs), knew about radio and served in many homes as the radio wizard. In many other homes, it was a young boy who assumed the role.

Radio listening today is mostly a companion to activities like reading, working, even jogging and is a passive pastime. In the early and middle 1920s, however, listening to radio was an activity in and of itself. It was not so much that radios were cranky creatures in need of perpetual prodding—though many were—but that for most people the joy of radio then was in capturing from the ether ever more feeble signals from ever more exotic places. In effect, the radio craze of the early twenties was amateur radio writ large (however, without transmitters).

The enthusiast usually sat transfixed before the radio apparatus, ears enclosed by phones, fingers slowly turning the knobs. Once a promising station was distinguished in the cacophony of hisses and squeals, other controls were adjusted to improve reception. This could take some time (some one-tubers had nine knobs). The radio maniac listened until—perhaps what seemed an eternity—an announcer finally disclosed the station's call letters and location. These were dutifully recorded on a log. Ironically, it was not the program that was of interest but the station identification. Sometimes the station faded out or was overwhelmed by static before its identity could be learned, and so a new search was begun. This activity is called Dx, the search for distant stations.

Radio Broadcast ran "How Far" contests, rewarding especially talented (and lucky) Dxers with new equipment. In July of 1924, Eva L. Rhodes, of Utica, New York, won first prize in the "Ready Made Receivers" division for an aggregate distance of 85,510 miles. She had logged 140 different stations, including one 2,480 miles away. This was real radio.

Howard Vincent O'Brien reflected on the Dx enterprise in *Collier's* after breathing life into his first radio, a one-tuber:

> I snapped the switch and out of this maze of wires of my own contriving came the soft Southern voice of Atlanta!
>
> I do not know what the soft voice said, because I was immediately seeking "something else." Some day, perhaps, I shall take an interest in radio programs. But at my present stage they are merely the tedium between call letters. To me no sounds are sweeter than "this is Station Soandso."
>
> In radio, it is not the *substance* of communication without wires, but the *fact* of it that enthralls. It is a sport, in which your wits, learning, and resourcefulness are matched against the endless perversity of the elements.

And, if the Dxer did stop long enough to listen to a program, what might have been heard in the 1923–1925 period? From WOK in Newark, in the early morning hours, was Bernarr Macfadden's calisthenics program during which exercises alternated with music. KFKB in Milford, Kansas, broadcast the lectures of one John Romulus Brinkley—alias Dr. Brinkley—who plugged his patent medicine potions and goat gland operation for restoring virility. If WOR, from Newark, New Jersey, had been extracted from the ether, then the uncompromising and often incisive political commentary of H. V. Kaltenborn might be heard. On Sunday mornings, services of the First Baptist Church were audible on WCBQ of Nashville, Tennessee. Chicago's WGN, spending $1,000 daily on cable charges, made available the last scenes of the Scope's trial. In addition, farmers got market quotations on various commodities, which meant that they could contemplate selling their produce beyond the local community at higher prices.

Although the roots of later broadcasting triumphs lie firmly within this (and the preceding) era, the radio experience of the early twenties was unlike that of Radio's Golden Age—the thirties and forties. For most

So little to do—such great results

GENERAL ELECTRIC

5.10   General Electric Tungar battery charger

5.11   Orchestra in a radio studio, ca. 1923

Americans, listening to particular programs was not yet an essential part of everyday life.

The programming ideal in the early twenties was to use live talent—either in the studio or piped in from a remote location. (Most of the record companies at first would not allow their discs to be played over the air.) Sometimes the "talent" was little more than the announcer—or a member of his family—singing or playing a musical instrument. Increasingly, though, stations were able to secure top-flight singers and musicians (Fig. 5.11). Even so, announcers had to be versatile in filling time. Phillip Carlin, of WEAF in New York, had on hand a book of poetry, which he could turn to in emergencies. One of the most successful announcers was Norman Brokenshire of New York's WJZ, whose improvisational skills were legendary. When several acts failed to show up in one day, Brokenshire sang, played the ukulele, played the piano, and in desperation hung the microphone out of the studio window to give his listeners "the sounds of New York." Judging by the huge amount of fan mail that a few early announcers received, we can be sure that some people paused long enough in their search for stations to become enraptured with certain radio personalities. Like the first movie stars, these announcers became celebrities—and demanded higher salaries.

Although many fine musical programs were aired in the early and middle twenties, including the nation's best opera companies and symphony orchestras, the quality of daily radio fare was highly uneven. In addition, programming was rather unpredictable. Listening at a regular time each day or week to one's favorite program did not become truly feasible until the very end of the decade. Radio sets and radio broadcasting would have to undergo many changes before the new medium could become, as Sarnoff envisioned, a household utility that everyone would want *and need*. But first there was the portable "boom."

6

Birth of the Boom Box

WHEN EDWIN ARMSTRONG STROLLED OFF THE ELEVATOR
with his proto–boom box blaring, RCA officials were impressed because
it played so well—for a portable. In 1923, the portable radio was already
a genre of commercial sets; Owen D. Young had certainly seen portables
before, for RCA was already making them.

The year 1923 does mark a watershed of sorts in the evolution of the
portable radio; that's when the first portable boom began. Actually, it
was less a boom than a boomlet, and in 1925 it went bust. Though short-
lived, the flurry of interest in portables included the arrival of the first
commercial boom box.

The cultural imperative of the portable radio was nurtured through
the late teens by a constituency of amateurs. Their home-brew sets continued to win prizes in radio magazine contests and now often incorporated vacuum tubes. However, crystal sets were not yet passé. In 1918,
one Audley V. H. Walsh presented his plans in *Electrical Experimenter*
for a "Vest Pocket" crystal set, which could even be tuned. The next year,
a "pocket size receiving set," built in a camera carrying case, was described by Joseph E. Aiken in Gernsback's *Radio Amateur News* (Fig.
6.1, top left). In 1920, *Radio Amateur News* offered a $100 prize for the
"Smallest Portable Radio Outfit." Though there were few entries, the
winning sets were elegant (Fig. 6.1, bottom left and above). Portables
were also thrust before the general public in war news, and, doubtless,
by returning soldiers extolling the virtues of portable wireless. Many
Americans were aware that portable radios had played an important role
in the war. They would also play a role, at home, in peacetime.

The automobile was another reason for the portable radio's expanding
constituency. In 1920, nine million cars were on the road; by the end of
the decade, nearly all middle-class (and many working-class) families
would own one. Throughout the twenties, automobile excursions and
tourism became more popular, as people took to their wheels on week-
ends and in the summertime. But, as of that time, paid vacations were
still restricted to white-collar workers, and so longer motor trips were
uncommon among much of the middle and working classes.

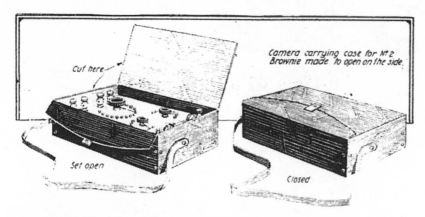

6.1   Small home-built portables of 1919–20. *Top left,* Joseph E. Aiken's pocket receiver built in a camera case, 1919; *bottom left and above,* two views of J. Blanchard Armstrong's prize-winning home-brew portable, 1920

Motels and roadside diners sprang up along major highways to serve the many Americans who had begun their love affair with the automobile. Accompanying the leisure seekers came an assortment of new portable products, including suitcases, picnic baskets, and portable phonographs (Fig. 6.2). Sales of the latter surged in the early twenties with the spread of automobile excursions; they were offered in dozens of models at $20 to $50. Ads for portable phonographs were targeted at trendsetting young men who wished to have, along with their new car, the most appropriate accompaniments to impress—and woo—young women.

It was a simple matter for radio makers to associate outdoor radios—portables—with leisure and good times. Photographs in radio magazines showed sets of all sorts being used outside, with young adults apparently enjoying themselves. For example, in a 1922 issue of *Wireless Age,* one Reta Pam was pictured on a golf course with her golf bag and tiny radio, apparently "catching music" between holes. Especially frequent were scenes of a radio on board a canoe or motorboat, entrancing its occupants (Fig. 6.3, above). Pictures of radios on camping trips were also common. Clearly, there was every reason to expect the portable radio to be as popular with the mobile set as the portable phonograph. The humorous side of outdoor and mobile radios was also explored (Fig. 6.3, below).

6.2   Outing portable phonograph, 1923

6.3   Outdoor radio is a subject of magazine art and humor at the beginning of the radio craze. *Above,* radio on a double date, 1923; *below,* a cover of *Radio News,* September 1921.

Shortly after the advent of commercial broadcasting, a few companies began to manufacture radios advertised as portables. Because all tube radios of that time were battery-powered, in principle (and in practice) they could be set up and operated anywhere. Most battery sets, though, did not have handles, rugged carrying cases, built-in speakers and battery compartments, and other amenities that enhanced portability. These features gradually developed in the "portable" radios made commercially from 1921 to 1925, as the portable underwent a transition from a battery set that could be used as a portable to one that was designed to be portable.

Only a very few commercial portables were put on the market prior to 1923. One of the earliest, which actually preceded KDKA, was the Winn "Portable Wireless Set No. 149," advertised in mid-1920 (Fig. 6.4, above). It was a combination receiver-transmitter that employed a crystal detector and spark transmitter—not exactly state-of-the-art radio technology. Housed in a "handsome mahogany" box, the set was obviously aimed at the amateur. Complete with phones, it fetched $50.

A year later, Grebe offered its Model KT-1, also a receiver-transmitter, this one with tubes (Fig. 6.4, left). Amateurs were the intended market, and it was said to be "Especially Adaptable and Efficient for Boy Scouts and Tourists." Its portability-enhancing features included a carrying case and compartments for batteries and headphones. Though it did not have an internal antenna, the radio was said to perform well with the "Grebe Radio Kite." The set sold for $175 and was apparently short-lived.

About the same time, Westinghouse was selling a crystal set, the Aeriola, Jr., which was, the ads assured, "just the outfit to take on camping trips or automobile tours to keep in touch with your home town." This radio was built into a small wooden box and weighed five pounds. (Many people believe that the Aeriola, Jr., was the first commercial set designed specifically for reception of commercial broadcasts.) A similar radio was RCA's Radiola 1. Ads boasted that the Radiola 1, which debuted

6.4 Ads for commercial portable radios, 1920–21. *Above,* Winn Portable Wireless, No. 149, 1920; *left,* Grebe KT-1, 1921

6.5 NBS demonstration portable sets. *Above,* early NBS portable with horn speaker, *1920; below,* Francis Dunmore posing with the NBS "Radio Valise," a completely self-contained portable, ca. 1921

in 1922, could be "opened like a book" and "carried like a satchel," but an antenna and ground were still needed at the end of the journey.

Like Armstrong's mobile superheterodyne, the most important portable of this era was a one-of-a-kind set that was used by its makers for technological display. The radio was made by engineers at the National Bureau of Standards (NBS). Founded in 1901, the NBS is a series of U.S. government laboratories that does high-powered but very esoteric science, often for other federal agencies. Although its work is highly regarded in the scientific community, few NBS projects have received much attention in the popular media. Occasionally, however, NBS people would build and prominently display an unusual (that is, newsworthy) product; in a few cases these have been portable radios of "advanced" design. Such diversions showed that NBS research did have practical applications and, especially, that its personnel were exceedingly clever.

After the war, the NBS was experimenting with radio direction finders. A few of their test receivers were installed in suitcases fitted with horn speakers, which made them eminently suitable for technological display (Fig. 6.5, above). One engineer working on this project was Francis Dunmore, whose amateur station had been featured in *Modern Electrics* back in 1909 (Fig. 4.2, left)—the issue before Miller's suitcase set (Fig. 4.5, above). One day in the spring of 1922, Dunmore was waiting in line to register at the Drake Hotel in Chicago. Suddenly, other patrons in the lobby heard the sounds of music from a source unseen. Finally, a few people noticed that Dunmore's suitcase was not quite normal; a crowd gathered, wanting to know if it was haunted. He got somewhat different questions when he demonstrated the "singing valise" at a meeting of the American Institute of Electrical Engineers.

The NBS portable radio was completely self-contained (Fig. 6.5, below). The external antenna and ground had been replaced by a "loop" antenna—a wire 50 to 100 feet long—wound in a coil that attached to the antenna and ground terminals of the radio. This loop antenna was enclosed in the case along with the batteries and a wooden horn speaker. According to a Signal Corps pamphlet (prepared by the NBS), the portable could receive "good signals from near-by stations." Gernsback's *Radio News* regarded this six-tube receiver as "perhaps one of the first portable sets ever constructed."

In 1921 and 1922 the "singing valise" brought Dunmore and the NBS favorable publicity in the press; more importantly, though, this forward-looking design established a standard for the portable radio that manufacturers could aim for. It is convenient to regard such a completely self-contained set as a "true" portable, that is, one that approaches the modern definition. It is fitting that the NBS "singing valise" now resides at the Smithsonian.

One of David Sarnoff's most intriguing visions also took shape in 1922, and concerned the portable radio. In a letter to A. N. Goldsmith, RCA's director of research, Sarnoff claimed that the next stage of radio would involve "developing a suitable receiver in such compact and efficient form as to enable an individual to conveniently carry it on his or her person." Such a radio, he continued, should approach as an ideal "the watch carried by a lady or a gentleman, which is not only serviceable but ornamental as well." He called his Lilliputian set a "Radiolette," and suggested that it could also function as a flashlight. In his reply to Sarnoff, Goldsmith affirmed the desirability and ultimate feasibility of personal radios but insisted that his laboratory would need more money to develop the Radiolette. Though RCA soon made true portable radios, apparently no effort was dedicated to miniaturization; their smallest set was about the size of a large mantle clock.

By the end of 1922 several companies had exhibited at radio shows portables with built-in batteries and loop antenna, including Radio Guild and Radio Units (Fig. 6.6c, e). However, these sets were not advertised nationally, and possibly were not even marketed. The Oard Phantom Receptor was put on sale, in late 1922, and it too had a number of portability-enhancing features (Fig. 6.6a), as did the Q-T portable (Fig. 6.6d).

b

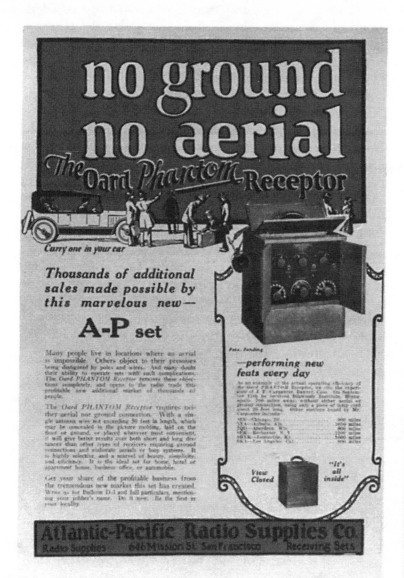

a

c

d

Crosley in early 1923 offered its "Model VI Portable," a small suitcase set with two tubes, completely self-contained except for the antenna. In an ad in *Wireless Age* it was shown hooked up to a loop antenna, but no such antenna was available as an accessory. The price? $40 "without batteries, tubes, and phones." A portable version of the larger (three-tube) Crosley Model VIII was also offered for $60. The Whiteland portable (Fig. 6.6b), also from early 1923, had a built-in speaker but, like the Crosley, lacked a loop antenna.

About the same time the A-DE-CO Radio Company was selling an assortment of loop antennas, which could "Make Your Set Portable for Camping and Vacation." Called the Warren Radio Loop, they ranged from $12 to $25.

Radio makers had every reason to believe in 1923 and 1924 that the potential market for portables was large. The magazines read by radio fans and radio makers alike continued to provide evidence in every issue that there was interest in portables. One of the most common pictures from that time is a radio being used outside the home or in a car (Fig. 6.7a, b). For example, the February 1923 *Radio Broadcast* pictured boy scouts from Long Island setting up a radio receiver in the woods (Fig. 6.7c). Later in that same year, it was reported that the New York City Police Department had put radios on some of its automobiles and motorcycles (Fig. 6.7e). *National Geographic*, also in 1923, published an article about America's marvelous passenger train system (it *was* marvelous then), including a photograph of radio apparatus and happy listeners that noted how "some of the progressive railroads are installing radio receiving stations on their big through expresses." Exploring expeditions of the early 1920s also took along radio receivers, sometimes to carry out experiments in reception, and these radio-accompanied expeditions were well reported in magazines and newspapers. Radios were pictured as well in baseball parks, city parks, national parks, and horse racing parks (Fig. 6.7d).

Lending credibility to the belief that portables could figure in outdoor leisure activities, especially for people who owned cars and had ample leisure time, was the picture of Edwin Armstrong, famous designer of radio circuits, at Palm Beach. Nattily attired in coat, tie, and hat, with the surf pounding immediately to his rear, Armstrong held a huge portable radio by its handle. The set was installed in a suitcase, sporting a horn speaker that emerged almost completely from the case near the handle. Armstrong had presented the home-brew portable—a superheterodyne, naturally—to his wife for their honeymoon.

Prior to mid-1923, the receivers pictured outdoors were mostly ordinary production models that were toted, batteries and all, to sometimes exotic places. Radio makers doubtless saw these pictures and envisioned new markets for radios that were *designed* to be carried about more easily. However, producing a good portable radio would require additional research and development.

The tube portables of 1920 to 1923 were not outstanding performers. None of these early commercial sets had the five or six tubes that could give excellent performance like the NBS "singing valise." The number of tubes had been kept to a minimum to make operation feasible with dry-

e

6.6 Several commercial portables of 1922 and early 1923. *a,* Oard Phantom receptor, 1922; *b,* Whiteland portable, 1923; *c,* Radio Guild portable, 1922; *d,* Q-T portable; *e,* Radio Units portable, 1922

a

b

c

d

e

6.7 Interest in outdoor radios continues to build in 1923–24. *a,* "Radio Ads to the Joys of the Hunt"; *b,* a car radio; *c,* Boy Scouts fiddle with receiver and huge loop antenna; *d,* jockeys tune in at Belmont Park; *e,* radio-equipped police motorcycle, New York

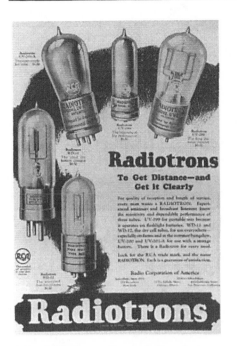

6.8   Ad for RCA tubes, including the UV-199, *top center*

cell batteries. The available tubes, such as the WD11, consumed quite a bit of power, and so battery life with five or six tubes would have been frustratingly short.

These problems were well known and required a creative technical solution; obviously, a low-drain tube was needed to supply the anticipated demand for good portable radios. General Electric began serious work on such a tube in 1921, achieving success the following year. In late spring of 1923, their new tube—UV-199—became available to radio makers; it consumed only .18 watt (compared to the WD11's .275 watt). A six-tube set could play two hours a day for three weeks before the A battery (three ordinary dry cells) wore out. Here was a tube around which powerful portables could be built. At about one inch in diameter and a little more than three inches long, the UV-199 was a little smaller than other RCA tubes (Fig. 6.8); clearly, though, it had not been miniaturized in the expectation of building Sarnoff's "Radiolettes." The UV-199 was announced at the New York Radio Show in 1922, and it stimulated a flurry of development activity among radio manufacturers.

RCA, of course, brought out a new portable as soon as the tubes became available in mid-1923. The Radiola II (Fig. 6.9), a two-tube regenerative set, was explicitly advertised as a "Portable Radiola." It cost a princely $97.50. Although mostly self-contained, the Radiola II still needed an external antenna. How their ad enticed: "Take it along—up to the mountains—out to the beach. On every auto trip and every camping trip." The portable radio was now clearly being pushed as a perfect companion to pleasurable outdoor activities.

With a powerful radio station nearby, the Radiola II could actually drive a speaker. Publicity photos printed by newspapers and radio magazines show the Radiola II with horn speaker animating a merry group—one man and four women—on a Florida Beach. The male viewers of this picture would have been expected to conclude that the set belonged to the man (who was closest to it), and that it was an efficacious lure for attracting comely persons of the opposite sex.

6.9   RCA's Radiola II being used during the National Balloon Race, ca. 1923

Among the new sets of 1923 was the Operadio 2, which some authorities believe was the first modern portable radio. Alan S. Douglas has pieced together the history of the Operadio Company and its portables in *Radio Manufacturers of the 1920s*. The Operadio 2 was manufactured by a tiny Chicago firm founded in 1922 by J. McWilliams Stone. In the pre-broadcasting era, Stone had been an amateur and took pride in having built portables—while a child—as early as 1908. His first commercial product was the Operadio 2, a path-breaking portable that was completely self-contained (Fig. 6.10, above); batteries, horn speaker, and the rest were all housed in a "compact carrying case." The loop antenna was placed in a detachable lid and front panel of the case. When the listener was ready to tune in, the case-antenna was removed and fixed to the top of the radio where it could swivel. With six tubes, the Operadio 2 was, for its time (late 1923), an impressive performer. However, at $190 (with tubes and batteries), the radio was unequivocally a luxury item.

Though not nationally advertised, the Operadio 2 was still heavily promoted. Stone was adept at generating publicity, and meticulously posed shots of the Operadio 2 (and later models) "in use" were printed in newspapers and radio magazines. The 1925 Operadio was lavishly advertised in radio magazines. To stress its ease of operation—it had vastly simplified controls—the set was often shown with a woman (Fig. 6.10, right).

Although very portable, the Operadio 2 still required some set-up time (thirty seconds said an ad), and, once set up, this portable fast lost portability, since it could not be played at all while being carried. It was not a boom box. Nonetheless, the Operadio 2 and later "improved" models—all pretty much the same electronically—enjoyed success for several years as the first stand-alone portable.

Another firm in Chicago, Zenith, also is given credit by many experts for making the first modern portable. Like Operadio, Zenith had roots deep in the amateur era that preceded World War I. One of Zenith's founders, R.H.G. Mathews, began selling radio apparatus to other amateurs in 1915. After the war he was joined by Karl Hassel and together they established the Chicago Radio Laboratory in 1918. To distinguish their products, the pair adopted the trademark Z-Nith (which derived from the call letters of their amateur radio station, 9ZN).

Z-Nith products were selling well to amateurs. When broadcasting mushroomed in late 1920, however, the company lacked capital to expand production. By happenstance, one of their customers, Eugene F. McDonald, Jr., was in a position to help out. Though not an amateur, he was intrigued enough by the new radio phenomenon to seek out a set at the Chicago Radio Laboratory. McDonald had accumulated some wealth before the war, having established a very successful automobile finance company, and in early 1921 was looking to start up a business. McDonald's interests and abilities and the Chicago Radio Laboratory's needs were a perfect match. In 1923 McDonald founded the Zenith Radio Corporation, with himself as president, to market the products of the Chicago Radio Laboratory; soon the companies merged completely. Not only did McDonald have business experience, but like so many Americans born in the late nineteenth century, he was a tinkerer, intrigued by

6.10 Operadio portables. *Above,* drawing of Operadio 2; *right,* 1925 Operadio and early film star Estelle Taylor Dempsey

electrical and mechanical things. He took a keen interest in product development and contributed many useful ideas. Above all, McDonald was a wealthy, trendsetting consumer who had insights into what products might have market appeal. But he was no Henry Ford; it mattered not whether Zenith workers could afford Zenith radios. Above all, Zenith would aim for the high-end market and sell only high-quality products. Indeed, Zenith did become an elite brand that inspired an unusual degree of customer loyalty.

McDonald himself became fabulously wealthy, living in a 185-foot yacht moored on Lake Michigan in Chicago. An explorer and adventurer, he pursued a glamorous lifestyle—his parties were legendary—that provided inspiration for many Zenith products.

Because of his Navy ties (he was a lieutenant-commander) and love of travel, long-distance communication was one of his interests. In collaboration with Admiral Donald B. MacMillan, famous Arctic explorer, McDonald convinced the navy that short-wave apparatus, heretofore the province of amateurs, would serve their needs. They clinched the argument in 1923 with a two-way communication between Greenland and Tasmania—a distance of 12,000 miles.

For two decades, Admiral MacMillan hauled Zenith prototype radios around the world, testing their reception under realistic conditions. Photographs from MacMillan expeditions and testimonials were frequent in Zenith ads.

In 1923 MacMillan took along to the Arctic an early version of a new Zenith radio—their first portable (Fig. 6.11, below). Apparently, it passed the field tests and went into production in the spring of 1924 (Fig. 6.11, above). In national advertising that began in April 1924, Zenith's "Companion" (occasionally called the "Superportable") was said to be "MacMillan's choice."

6.11   The Zenith Companion, 1924. *Above,* interior of Zenith Companion, perhaps a prototype; *below,* Admiral Donald B. MacMillan with Zenith Companion

The Zenith Companion was, finally, the portable that had it all, as their ads proudly proclaimed: "Here's a six-tube radio set that's entirely self-contained—tubes, A batteries, B batteries, loud speaker and loop antenna complete." Although of briefcase design, the receiver could be played with the case closed because it had two controls on the exterior, near the handle. Ads show this radio in a variety of settings, indoors and outdoors.

Like the rest of the Zenith line, the Companion was expensive—a staggering $230 complete—and heavy (24 lbs). It had six tubes and employed a circuit that compromised sensitivity and selectivity in favor of one-knob tuning. The set was on the market for less than a year; apparently its performance as a radio fell below expectations—MacMillan's testimonial notwithstanding. In one story (which I could not verify), Zenith is said to have recalled the few hundred Companions that had been sold, destroying them and refunding the purchase price. Another mystery shrouds the Zenith Companion. Within Zenith today, only one example of an early portable survives. It is believed to have been McDonald's personal property. When I examined this set at Zenith's headquarters in 1989, I found that it was not a Zenith at all. A small nameplate identified this briefcase portable as a "Westburr Six." The

Westburr Six is a radio of uncommon obscurity, but some information did turn up about the set and the company that made it.

Westburr was a small firm that received authority to do business in New York State in December of 1923. Apparently, its only product was the briefcase portable. The first public notice of the Westburr Six appeared in *The Radio Dealer* in February, 1924—two months before the earliest ad for the Zenith Companion. The set was described, illustrated (Fig. 6.12, above), and was said to have been "recently placed on [the] market." A second (and briefer) blurb accompanied a picture of the set's insides, and was published by *Radio Age* in April 1924 (Fig. 6.12, right). The radio was also given a favorable writeup in the *New York World*.

Sold mainly in the New York area, the Westburr Six was not advertised in radio magazines. Two ads did appear in the *New York Sun* radio section; the first (April 19, 1924) included the price ($170—less tubes and batteries), while the second (November 8, 1924) announced a close-out of the 1924 model (Landay's, a department store that reputedly was a partner in Westburr, had the sets on sale for $79). It is easy to infer that the Westburr Six, like the Zenith Companion, was not a hot seller. Apparently, though, McDonald at Zenith bought one.

The Westburr Six may have been on the market a few months before the Zenith Companion, and so a case can be made that it was the first boom box. However, there is no doubt that Zenith's Companion was the first boom box advertised nationally. Whether it or the Westburr Six or the Operadio 2 was the first essentially modern portable radio is mostly a question of definition—about which radio collectors and historians will argue for some time to come. In view of the social forces present in the United States in the early twenties (in particular, the increasing use of the car for recreation) and their influence on radio technology (development of the UV-199 tube), experimentation with portables was inevitable. Sooner or later a company would have followed the NBS's lead and cre-

6.12 The Westburr Six, 1924. *Right,* Miss Claire Patton showing interior view; *above,* exterior view

ated a completely self-contained set; it was, after all, a cultural imperative of some antiquity that had achieved the state of feasibility.

Interest in portables intensified over the next two years, with many models released. A panorama of the new portables is shown in Figure 6.13. The "Pocket Radio" (Fig. 6.13c), made by Auto Indicator Co. of Grand Rapids, Michigan, was the smallest tube set produced commercially in this era (though still too large for pockets).

To celebrate the portable's apparent coming of age, *Radio News* in August 1925 published Gernsback's "First Annual Portable Radio Set Directory," which listed twenty-two models from sixteen manufacturers. Included were RCA's Radiola 24 (Fig. 6.14, above) and Radiola 26 (Fig. 6.14, left); not only were these two models self-contained, but they were also superheterodynes, offspring of Armstrong's original.

d

j

k

l

m

n

o

6.13 The Portable Boom of 1924–25: A panorama of commercial sets. *a*, Karryadio; *b*, Wells Bear-Cat 500; *c*, Auto Indicator Pocket Radio; *d*, Crosley 51-P; *e*, Haynes-Griffin Superheterodyne; *f*, The Portable Voceleste, General Radio Manufacturing Corp., New York; *g*, Ozarka; *h*, Kodel P-11; *i*, Hetrola R-199-P; *j*, Echophone; *k*, Standard Portable; *l*, Telmaco Acme; *m*, Cutting and Washington 12A; *n*, Kodel P-12; *o*, American Apparatus CN-7; *p*, Kennedy Model III

p

There is no loneliness where there is a Radiola.....

6.14 RCA's self-contained portables of 1925. *Above,* the Radiola 24 is shown in this 1925 RCA publicity photo; *left,* ad for RCA Radiola 26

Designs varied greatly among the twenty-two models. The buyer had a choice of one to eight tubes; a few sets had built-in horn speakers, though most did not; controls ranged from two to six; and weight and size spanned the extremes, from the almost diminutive Kodel P-11 (Fig. 6.13h), three pounds, 6-1/4" × 9" × 5", to the Haynes-Griffin Super-heterodyne (Fig. 6.13e), forty-four pounds, 21" × 8-1/2" × 17-1/4". For the most part, however, these portables tended to be heavyweights: twelve sets exceeded twenty-five pounds. Much of this weight, of course, was batteries. Lifting thirty-five pounds with one arm is not so difficult, but balancing it comfortably while walking is more of a challenge. Clearly, the larger, better performing portables of 1924–25 could be carted around but, like today's big boom boxes, they were not carried for long distances "easily."

Portables also came in a range of prices. The two least expensive were the Crosley 50P and the Kodel P-11 (Fig. 6.13h), both at $16; they were one-tubers heard through headphones. The cheapest set with a built-in speaker was the Kodel P-14 ($50 plus accessories), but it lacked a loop antenna. Completely self-contained portables cost a great deal more: Operadio 1925 ($189 ready to play), Radio Masterpiece P.L. 103 ($160 without accessories), or the RCA Radiola 24 ($195 less batteries). The average price of a 1925 portable was around $100.

Regardless of size or price, most commercial portables had cases made of fabric-covered wood; a few, such as the RCA Radiola 26, were of varnished wood. At home, then, a portable was not apt to be regarded as a stylish knickknack or piece of furniture. Like the majority of home radios, the portable was an undisguised electrical apparatus. When closed up, most looked like luggage, a box, or maybe a squarish doctor's bag; hardly an item that would appeal to interior decorators. Thus, when the portable was resting at home from its travels and outings, it was likely to be out of sight; if used at all it was probably in a bedroom, den, or basement.

By the middle of 1925, the design of the portable radio had reached a state of maturity using the electronic technology of that age. With its smaller horn speaker, the portable sounded a bit tinnier than its parlor counterpart, but the best ones played reasonably well—for a portable. Portability had been achieved in the sense that the radios were self-contained and could be carried, occasionally with ease, to and from one's automobile.

The possibility that portable radios could someday be made small enough to carry around on one's person—perhaps even in a vest pocket—was still being explored in the early and middle twenties by experimenters. *The Radio Amateur's Handbook* of 1922 illustrated two minute crystal sets, one in a matchbox, the other mounted on a pinkie ring (Fig. 6.15c). The latter used an umbrella for the antenna. Still others were built in billfolds or carried in handbags (Fig. 6.15d, e). Tinkerers, of course, continued to build watch-case and watch-size radios (Fig. 6.15b). Periodically, *Radio News* ran pictures of oddball sets—the Radio Garter, radio-in-a-book, and so on—under the caption "Freak Receiving Sets." Needless to say, the bicycle radio made another appearance (Fig. 6.15a).

a

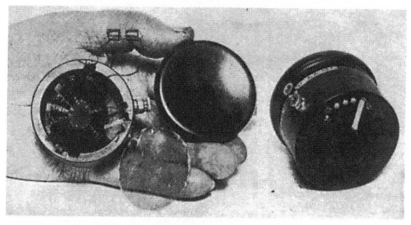

b

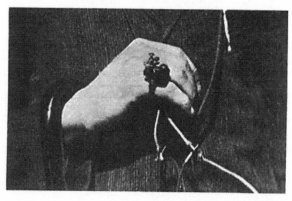

c

d

e

6.15 Some home-brew miniatures and novelty portables of the early twenties. *a*, James Scull shows off his bicycle radio, 1924; *b*, "A Radio Receiver the Size of a Watch," built by St. Louis jeweler J. A. Key in 1923; *c*, Alfred G. Rinehart's pinkie ring crystal set, 1922; *d*, Mr. Moore's billfold radio, 1921; *e*, radio in a purse, 1922

The pocket portable radio remained a cultural imperative—at least among radio enthusiasts. Indeed, Hugo Gernsback made the pocket radio seem old hat already in 1926. In introducing readers of *Amazing Stories* to H. G. Wells' delightful tale, the "Crystal Egg," about communication with Mars through a crystal egg, he boasted that "we who are accustomed to radio and who can bring voices out of the thin air with a pocket radio receptor, will not think that the crystal egg is impossible." And, indeed, a few companies did sell pocket-size receivers in the early twenties (Fig. 6.16). However, given off-the-shelf radio technology of that time, only crystal sets could be made pocket size; such sets were suitable only for children or other beginners in radio (see Chapter 11). A true pocket portable was still some years away.

In an even greater leap beyond portable radio practicability, Gernsback in his *Science and Invention* magazine envisioned the "sonophone." Using the vehicle of a dream ("Alice in Sound-land"), he introduced Alice to a pair of headphones that could make immediately audible electromagnetic waves of any frequency—including light and x-rays. One of Gernsback's more sinister visions, published a month after the sonophone, was the "Radio Police Automaton," a veritable robocop that included tear gas, a loud speaker, and "rotating discs which carry lead balls on flexible leads" for use as clubs. For "night attack" the radio-controlled, bullet-proof automaton (the word "robot" was not around yet) had headlights. These machines would be valuable, Gernsback assured his readers, for dispersing mobs (perhaps those gathering to watch Coolidge snooze?).

Though there seemed to be no limit to Gernsback's technological imaginings, most of his excursions to the future were based more-or-less on sound techno-mancy. In 1924, for example, he authored an editorial in *Science and Invention* about "Airplanes vs. airships." At that time, it was not yet evident that the dirigible's future was to be drastically limited. He argued that the airship had fundamental shortcomings, such as slow speed and high cost of operation, that would prove decisive in competition with the airplane. The latter, he predicted, would prevail as the main mode of air transportation.

Much of Gernsback's techno-mancy concerned electronic devices in the home. In mid-1925, for example, he prophesied that every house with a "first-class radio set" would soon own a "radio recording device" to capture favorite programs permanently. In December of that same year, *Radio News* illustrated a combination radio-recorder that employed not tape but magnetized wire.

Television was another of Gernsback's great loves, and his optimism knew no bounds. In May of 1926, he forecast in *Radio News* that television would be "perfected" within two years, with color coming later. Gernsback even foresaw the possibility that amateurs would communicate with television phones. In fact, some commercial televisions were sold in the late twenties, but no one seriously claimed that the technology had been perfected. A more typical assessment of television's future was articulated in *Radio Broadcast* the following year by Carl Dreher: "The everyday application of television is a remote possibility in five years, a fair possibility in ten, a probability in fifteen."

Children growing up in the twenties (and thirties) received in Gernsback's magazines and in other popular media a heavy dose of technological enthusiasm. The belief that technology could solve any problem—even social problems ("mob control")—had taken root in the American consciousness. It could not have been otherwise in a decade that saw so many of the earlier promises approaching fulfillment. The future obviously would bring a cornucopia of wonderful new things. Among the artifacts of tomorrow, of course, would be better portable radios.

In looking at the commercial portable radios of 1924 and 1925 from the vantage point of today, we can see much room for improvement. Indeed, we would expect that in the years immediately following 1925, sets would have been streamlined and upgraded somewhat: tubes miniaturized, speakers perfected, and better and lighter batteries built. Above all, we would expect the price of portables (around $200 for a completely self-contained set) to fall so that this luxury item could spread beyond the wealthy trendsetters.

Those who held off purchasing portables in 1925 in anticipation of improvements and price reductions were in for a surprise. By 1926, the number of portables being manufactured had dramatically dropped. Those remaining were unchanged—in price and performance. Hugo Gernsback's Portable Directory of 1925, billed as the "First Annual," was also the last. Portable sales failed to meet expectations, as Gernsback lamented in a *Radio News* editorial of July 1926:

> The surprising thing, however, is that, at least in this country, such sets [portables] have not attained the great popularity which they deserve.
>
> There is really no reason why every automobile, every motorboat, and, for that matter, every home or office, should not have its portable receiver, because its utility is paramount.
>
> While, of course, untold numbers of portable receivers have been constructed, either by private builders or by companies making sets commercially, it is surprising how few such sets are actually in use.

The portable "boom" had ended.

The following month *Radio News* published an article by A. P. Peck entitled "Are portable sets really practical?" The conclusion, of course, was "emphatically, Yes!" Though intended to spur portable sales, the article tacitly reinforced the view that portables were still a pain. Nearly half the piece discussed how to improvise antennas and grounds in the wilderness. Actually, these problems stemmed in part from the limitations of broadcasting at that time, which put a severe handicap on portable radios. Because most stations had transmitters of 500 watts or less, reception beyond twenty or so miles was iffy at best, especially in the daytime. That is why even the better sets needed good external antennas to pick up anything in the country. In any event, it was probably too late to reverse the commercial portable's fate. By 1927, scarcely anyone was advertising portables.

The death of the portable was nearly complete, and it took some companies with it. Westburr was one of the first to expire, forced into bankruptcy in early 1925. Other firms heavily into portables in 1924 and

a

6.16 Commercial pocket radios of the early twenties. *a,* R.P.C. Midget Pocket Receiver, 1923; *b,* Pal Radio, 1924; *c,* Garod Heliphone pocket radio, early twenties; *d,* Beaver Baby Grand "Vest Pocket Radio Receiving Set," 1922

b

c

d

1925, including Kodel, Armley (Karryadio), Telephone Maintenance Company (Telmaco), and Hetrola, were simply not making any radios after 1927. Operadio survived into 1927, then entered receivership. (It did revive, though, and still exists as the DuKane Company.) Of the sixteen set-makers listed in the 1925 portable directory, only four were still doing radio business in 1930. In sharp contrast were many of the most successful radio firms of the twenties, including Atwater Kent, which made no portables in that decade. Other prosperous companies (RCA, Zenith, Crosley) sold no portables in the late twenties.

The savvy firms had concluded by 1926 that manufacturing portable radios was suicidal. The larger companies had learned the lesson, perhaps too well, that portables did not appeal to the masses. Obviously, there was no point in perfecting portables, reasoned radio company executives, because Americans wouldn't buy them. Zenith's groundbreaking Companion had no successor in that company for a decade and a half. Philco, one of the most profitable radio firms of the 1930s, made none until late 1938. An early leader of the portable pack, RCA manufactured only one (in 1932) prior to the portable revival in 1939.

Apparently, the radio companies badly misjudged the market for portables. Many of the executives and engineers in these companies had, of course, been weaned on Gernsback's magazines, and were part of the portable radio constituency. It would be surprising had they done anything but draw the most optimistic inferences from the interest that people had shown in "summer" and "outdoor" radios prior to 1925.

Manufacturers erred in their belief that portable radios could play the same role in automobile outings as portable phonographs. In principle, both could supply music for good times outdoors; in practice, however, there were important differences (beyond the obvious one of purchase price). Not only was radio reception erratic in the country, but one could not count on finding a station with the right program at the right time. Imagine the chagrin a young man might have felt, trying to impress his date on a picnic, if the only sound that came wafting through the ether was the voice of Dr. Brinkley hawking his goat-gland operation. The phonograph could at least be counted on to provide music for the mood; the portable radio was an unreliable substitute. Of course, there were other uses for the portable radio, and a failure to perform one function does not doom a product to extinction.

Pictures of radios in use provide some clues to other functions of the portable radio in the early and middle twenties, but care is needed in their interpretation, since so many were posed. It appears that portable radios appealed mainly to men, especially wealthy men inclined to tinker. This should not be surprising to us because then—as now—Americans (mostly men) have been fascinated with novel things advertised to be on the cutting edge of technology. The makers of such gadgets are assured of a small market of people—the trendsetters—who can afford to experiment with new products to see if they are compatible with their lifestyle. In the meantime, the owner of a new item, showing it off and explaining its operation, becomes a center of attention and so earns prestige and admiration from friends and acquaintances for his new-found expertise.

The things we buy (like cars and radios) do more than move us from place to place or capture invisible waves from the ether; they make—it is almost trite now to note—*statements* about their owner. A Victrola ad from 1923 said it all: "People express themselves in their possessions." Using things, we silently convey information about our fashion-consciousness, special know-how, wealth and disposable income, social position, access to arcane knowledge, even availability to the opposite sex. These important social functions ensure that nearly any product, however "impractical" or expensive, will find at least a small market—*if it sends the right message.*

Items that survive this initial period of experimentation must have appeal beyond mere novelty. Friends become impressed enough to buy their own product, and its popularity increases. With the economies of mass production, the price soon drops, making it affordable to a much larger number of upper- and middle-class Americans. The portable radio of the 1920s did not make it past the stage of early experimentation. The appeal started and stopped with tinkerers of means who themselves became props in pictures of portables "in use." Few people came to believe

6.17 Central Park radio concert attracts avid listeners and a photographer, 1923

that radio had to be a constant companion. (This probably had less to do with the state of portables than with the general state of radio and, especially, broadcasting.)

By setting up a "portable" radio on a park bench or other unlikely spot, the owner could rapidly attract attention—and maybe a photographer (Fig. 6.17). However, the novelty of displaying and seeing radios in unexpected places wore off quickly, especially as radios of all sorts became more common, and such pictures were seldom published in radio and mass-circulation magazines after 1925. Wealthy tinkerers had available many new electronic marvels to occupy their leisure time and impress their acquaintances, such as hi-fi sets and television, and it was these items that captured their attention—and that of photographers and writers—in the late twenties and thirties.

Finally, the lifestyles of most Americans in the mid-twenties were not much affected by the new medium. This was obviously true for the majority of adults, who as yet owned no radios; but even the radio enthusiast, the die-hard Dxer, found in radio little more than a pleasurable evening pastime. Radio was not yet the source of daily news and "favorite" programs that it would become in the thirties. Not until broadcasting and the activities of ordinary people changed would there be a mass market for portables.

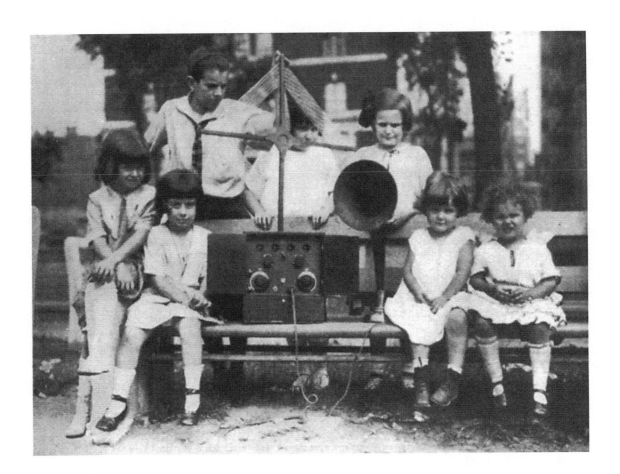

Although a limited technological success, portables of the early 1920s were for the most part a commercial failure. On rare occasions, however, a failure is spectacular. Consider, for example, the snub-nosed Chrysler "air-flow" car of 1934. This radical design anticipated concerns about aerodynamics and fuel economy that did not resurface until the seventies. But in the marketplace the air-flow was a flop. The wireless remote control of the 1930s, proclaimed by U.S. manufacturers to be "the most miraculous radio invention since radio itself," did not achieve popularity then.

Also, ahead of its time, the portable was to be a spectacular failure during the radio craze. Radio engineers encountered—and elegantly solved—the problems of portable design, putting a radio with its own speaker and batteries and antenna into a small box with a carrying handle. With its diminutive horn speaker, Zenith's portable would not have had a mellow tone, but in the evening it could have filled a tent—sometimes—with the sounds of the Barber of Seville or a boxing match. By the end of the decade, though, few people remembered the Zenith Companion.

7

Radio Gears Up for a Golden Age

DURING THE TWENTIES, SKEPTICS CARPED AND WROTE off radio as a fad that would soon pass. Thomas Edison, no less, had pronounced radio unfit for playing music. However, changes in American culture and in radio itself during the late twenties and early thirties ensured that the new medium would endure.

Life for ordinary Americans—the middle and working classes—had undergone dramatic changes, many for the better, since the time of Edward Bellamy. The idyllic society he envisioned, where goods were plentiful and available to all, no longer seemed so utterly fantastic in the twenties; lifestyles of the wealthy had become accessible to many. Though jarring, industrialization and urbanization had at long last brought about real improvement in standards of living.

Industrial capitalism worked its magic by dramatically raising both productive capacity and productivity: more goods were made with much less labor per item. These economies had two major effects on American lifestyles. First, the price of goods fell relative to the price of labor; this translated directly into an increase in real income and purchasing power. Second, the average number of hours in the workweek declined, leading to more leisure time.

In Bellamy's time, American working-class families spent more than 90 percent of their annual income (which was substantially augmented by child labor) on food, clothing, and shelter; needless to say, luxuries were uncommon. Few homes had bathrooms, indoor plumbing, electric lights, central heating, and the other amenities we now associate with civilization. By the twenties, however, many urban middle- and working-class families had these conveniences (which by then were regarded as necessities) as well as disposable income that could be used to buy a greater share of the "good life." Poverty was still widespread in the age of bathtub gin and flappers, but in only two generations America had become a country of mass consumption.

With shorter workdays and shorter workweeks, Americans could devote more leisure time to reading and other recreation. Magazines such as *Good Housekeeping* and the *Saturday Evening Post* targeted the growing corps of consumers with ads that introduced new products and helped to cultivate new needs. Describing the *Saturday Evening Post*, Frederick Lewis Allen in *The Big Change* clarified the role of the new print media:

> Through this five-cent magazine, and others like it, millions of Americans were getting a weekly or monthly inoculation in ways of living and of thinking that were middle class, or classless American . . . through the same media they were being introduced to the promised delights of autos . . . and breakfast foods that American industry was producing, not for the few, but for the many.

Ads no longer passively described a product, but pictured its beneficial effects on the buyer, often in lavish art. Department stores and mail-order catalogs also helped Americans to get the goods that could fill—or increase—their leisure time. Together these institutions disseminated brand names—8,500 by 1930—and created for them a nationwide consumer following.

Newspapers, magazines, and radio itself contributed to a news consciousness that extended beyond the confines of the ordinary activities of ordinary people. Diversion was provided by current events, from Lindbergh's solo flight across the Atlantic to the murder trial of Leopold and Loeb. More importantly, these happenings—and sports especially—became topics of conversation in the workplace where people, drawn from different families and different neighborhoods, might have almost nothing else in common. The new media helped to meet the insatiable demand for exotic trivia, contributing to the status of Babe Ruth and Jack Dempsey as culture heroes.

In the 1920s Americans were part of the world's largest culture of consumption. The new consumer society depended on purchases to keep factories humming, which in turn fattened capitalists' wallets and also gave workers the wherewithal to indulge. Though thrift and savings had once been important American virtues, new values and attitudes arose to promote purchases. As historian Robert McElvaine points out, "Old habits of thrift and sacrifice, deeply ingrained in early stages of industrialism, now were to be altered, if not reversed, and advertisers would be instructors in the new ways." The new morality of spending was epitomized in the social commentary of one Mrs. Bruère: "Nobody ought to enjoy doing without things, else we should become a race of misers, each sitting on his little separate store of gold." Quickly, it seems, many came to believe that spending was the road to prosperity. Though once reined in by lack of disposable income and puritanical restraints, the American family was being radically transformed from a center of production to a center of consumption.

Buying on the installment plan and annual changes in style are two features of consumer culture that we now take for granted, but they did not become common until the late twenties. The expansion of credit allowed for middle- and working-class consumers to buy cars, appliances, furniture, and even radios on "easy monthly (or weekly) payments." By the end of the twenties, installment purchases were responsible for 60 percent of all cars sold and 80 percent of all radios.

By the mid-twenties, industrial capitalism faced a new crisis, one whose solution would push the consumer culture to greater extremes. The problem was surplus production. American factories were capable of making far more products than consumers could use. New models with insignificant "improvements" might attract a few trendsetters, but the mass of consumers sat pat, content with their still-functioning older models, awaiting more substantial changes.

The problem first surfaced on a large scale in the automobile industry, as cars began to saturate the market. The Model T's price continued to fall, but so did demand because just about any family that wanted an automobile already had one—usually a black Model T. Even then, automobiles lasted many years, so replacement was going to be a slow process. Factories would be doomed to operate well below capacity, with profits down and workers laid off. Something had to be done to spur demand.

This time it was General Motors that had a better idea. In 1926 the auto giant brought out an entirely new Chevrolet, colorful and boldly styled, and for the first time outsold Ford. The Chevrolet triumph demonstrated that consumers beyond the trendsetters could be induced to replace a perfectly good product with one that had a more "modern" appearance. And so was born the doctrine of "stylistic obsolescence," which came to be applied to all consumer goods.

Stylistic obsolescence, of course, was not entirely new; it had become well established in the fashion industry in the nineteenth century. Mass-production of apparel had created the same problems of oversupply and declining demand, and the same solution had been found: annual changes based on style. In this way, manufacturers were able to exploit the consumers' constant quest for novelty to push products that no one really needed—except to conform to social expectations.

Once the doctrine of stylistic obsolescence was embraced by one manufacturer in an industry, sales successes stimulated the competition to follow suit. Soon other auto brands—even Fords—were undergoing a yearly face-lift, as manufacturers established "styling sections" where designers trained in art rather than engineering set the fashion trends. Other consumer-product manufacturers joined the styling movement in the late twenties and early thirties. At first designers drew mainly upon European styles and museum objects for inspiration; by the mid-thirties, though, a uniquely American "machine age" style had evolved. The strikingly new styles created by the first generation of "industrial designers" would be called, decades later, Art Deco.

As lifestyles altered during the twenties, the distinction between necessities and luxuries became totally blurred: In 1905 an automobile was clearly a luxury; by 1925 it was for many who had moved to the suburbs their only means of transportation—an obvious necessity. For radios to undergo the transition from luxury to necessity would require some big changes: receivers themselves would have to become easier to operate and acquire an improved appearance; perhaps more importantly, programming would have to become predictable and broaden its appeal. These changes did take place, in the space of just a few years, making radio the indispensable source of entertainment and information for the majority of American families by the mid-1930s.

As early as the mid-twenties the more savvy radio makers had begun to realize that one day soon every electrical tinkerer in America would own a radio. Though tinkerers and dedicated Dxers would continue to buy "new-and-improved" products, a much larger market awaited exploitation: families, with no special electronic expertise, that just wanted passive entertainment.

To reach this market, radios were dramatically redesigned—inside and out. On the inside they became more complex electronically. Battery eliminators were built in, ushering in the age of the "all electrics." This required a huge power transformer to supply A and B voltages, a rectifier tube (to convert the B supply from AC to DC), and filter capacitors to smooth out the last fluctuations in the B voltage. A new family of "AC tubes" had filaments that could operate directly on the transformer's alternating current. AC tubes rapidly became the mainstay of all consumer electronics—except for portables, which could use tubes that were powered by DC from batteries.

With these changes in tubes and power supply, only one power cord dangled from the radios' rear. To get the electronics functioning, one simply plugged that cord into a "light socket." The elimination of batteries and a tangle of battery cables was a simplification that everyone welcomed.

Although the all-electrics were more complex inside, innovations in tuning made them vastly easier to use. Tuning capacitors, previously operated individually, were "ganged" on the same shaft, and so could be adjusted with just one knob. Although a few sets had one-knob tuning earlier, it did not catch on industry-wide until 1927. One-knob tuning literally put radio within everyone's reach; even children and the infirm could easily seek stations. Tuning was no longer an arcane rite practiced by the family wizard.

A final feature of most all-electrics was a cone speaker. The horn, happily, was obsolete by 1928. Replacing it was a speaker in which sound was produced by the vibrations of a large parchment cone. Because cone speakers could reproduce a much wider range of musical notes than the ordinary horn, sounds emanating from a cone speaker—even a relatively small one—were startlingly realistic compared to the typical horn's tinny sounds. At once the standards of musical reproduction rose dramatically: radios of the late twenties sounded good.

With precious few (and costly) exceptions, receivers of the "radio craze" had been monstrosities. Though some radio cabinets were made of well-finished wood, even these sets—with their prominent dials and protruding knobs—seemed to say: "electrical device—stay away." And those batteries! They had to be hidden in a cabinet or under a table. By no stretch of the imagination could the ordinary radios of that era be regarded as furniture fit for the parlor or living room—despite frequent ads showing them in just those locations (Fig. 7.1).

7.1 This Eagle ad in *The House Beautiful* frames a not-so-elegant radio in a so-so-elegant setting, 1925

All of this changed, rather suddenly it seems, with the arrival of the 1927 and 1928 models. An article in *Radio Broadcast* neatly summed up the transformation: "The work of the artist and interior decorator—as well as the labors of the radio engineer—are evident . . . the radio receiver of 1927–28 is a thing of beauty." And so they were. From table top sets to imposing and ornate consoles, the new models had made the transition to furniture. On some sets, the controls were actually placed behind doors or panels. Many of the consoles had built-in speakers, a feature that would become standard by 1930. Even Crosley offered nice cabinets to house their homely metal-cased Bandbox (Fig. 7.2).

The employment of name designers, the use of expensive veneers, and the attention to style that became essential for radios of the late twenties reflected a change in marketing strategy. Radio makers had come to appreciate that women, who were thought to place stress on aesthetics over electronics, could influence radio purchases. Attractive sets would appeal to women, and so expand the market. In introducing pictures of the new models, *Radio Broadcast* was quite explicit about the new appeal:

When the Greatest Show *in* History thrills the World

"..... You're *there* with a Crosley"

$35

$85

The BANDBOX *$55*
A 6 tube Receiver
Brown frosted Crystalline finish—
Bronze Escutcheon.

$65

These approved cabinets have been selected by Powel Crosley, Jr., as ideal consoles, acoustically and mechanically, for the installation of the Crosley "BAND-BOX." Genuine Musicones built in. Crosley dealers secure them from their jobbers through

H. T. ROBERTS CO.,
914 S. Michigan Ave.,
Chicago, Ill.
Sales Agents for Approved Console Factories:

SHOWERS BROTHERS
COMPANY

THE WOLF MFG.
INDUSTRIES

A tremendous Crosley radio achievement for 1927-28

From the representative models shown on these pages, the housewife can gain an excellent idea of the appearance of moderate priced radio receivers which are offered to suit her taste as well as that of her husband. Her ideas of the necessary limitations of her domestic decorative scheme should blend with her husband's technical opinion.

Radio's metamorphosis did not come cheaply. Most of the sets listed by *Radio Broadcast* cost in excess of $200. One of the best-selling models of that time, RCA's Radiola 60, was $135 plus $35 for a matching speaker (Fig. 7.3). Though expensive, radios continued to be purchased by the millions each year. Radio sales reached over three-quarter billion dollars in 1929. However, more than half the homes in America still did not have radios.

In the last years of the decade, the pace of styling changes, heralded at the annual radio trade shows, became almost frantic. Increasingly, customers were enticed into buying expensive cabinets that happened to house radios. Style had taken command of the radio market.

As the all-electrics came into their own in the late 1920s, portables made a last stand. None of the major companies (RCA, Zenith, Atwater Kent, etc.) built them; instead, these sets were products of the entrepreneurial fringe, and several companies apparently succeeded for short periods. The Trav-Ler Company, in particular, offered a portable during every year from 1925 to 1930 (Fig. 7.4b); their sets had five tubes and ranged in price from $57.50 to $75.00. Kemper Radio also sold portables in the late twenties, but disappeared from the marketplace after 1929 (Fig. 7.4a). A third pre-Depression player in the portable game was Automatic Radio, which offered several "Tom Thumb" models (four and five tubers from $57.50 to $95.00) in 1929 and 1930 (Fig. 7.4c).

Electronically, the late twenties portables were little changed from those of the earlier portable boom. Yet, because transmitters were now much more powerful (some stations exceeded 100,000 watts), the same portable radio became a more reliable performer in the country. Even though portables were much cheaper and had achieved a new level of practicality (by virtue of the changes in broadcasting), there was no stampede to buy them. The reasons are many.

7.2 Even the cheapies get a facelift: a Crosley ad from 1927

7.3 Radiola 60 (*center*) in an RCA display, 1928

a

b

c

In the first place, the new companies did not promote portables in mass-circulation magazines. The portable was pitched to potential markets of limited size, and the sets were not especially well distributed through retail outlets.

Even more telling was their failure to appeal to the trendsetters. The late twenties portables did not participate in the styling craze that hit other home radios. The small companies that made these sets probably could not afford styling sections. Their homely handiwork would hardly be harbingers of portables to come; undistinguished and unnoticed, they were in no way capable of attracting trendsetters and leading a portable radio revival.

Fed by prosperity during the last roar of the twenties, these portable makers apparently catered for a short while to a small and highly specialized market. That the market was minuscule and shrinking is indicated by the failure of Trav-Ler, Automatic, and Kemper to offer any portables in the early 1930s.

Although portables were slowly disappearing from view, the cultural imperative of ultimate portability in radios was kept very much alive by the fraternity of electronic tinkerers. In April of 1928, for example, one J. B. Armstrong presented his design for "An Efficient Little One-Tube Receiver, Which Weighs Less than Four Pounds." Lamenting that most "portable" radios were about as portable as a steamer trunk, Armstrong offered his alternative, which "can be carried in one hand and will not cause drooping of the shoulders." His radio was built into an aluminum case that included batteries (Fig. 7.5). Although an external antenna and ground were needed, strong local stations could be heard "using a loop consisting of three feet of wire laid across the shoulders and hanging vertically." From the outside, Armstrong's compact portable, with carrying strap on one end, was about the shape and size of a camera. When interest in portables rebounded before the Second World War, camera-style portables would become an important genre.

Prior to 1927, the boundary between portables and nonportables was not always sharp, for all radios were powered by batteries. This is clear from the photographs of "portables" in use; many were not radios advertised and sold as portables. By contrast, when the all-electrics entered the scene in large numbers, beginning in 1927, the distinction between portables and nonportables became quite rigid. For all their good sound and ease of operation, the all-electrics could not be operated on batteries, and so new kinds of portable radios were needed.

Because portable radios continued to have important business and military uses in the late twenties and thirties, specialized portable transceivers were developed for these "mobile" applications. Radios for police, airplanes, and ships, for example, became available. Moreover, special frequencies outside the AM broadcast band were allocated for these transmissions.

Radio communication was also essential for many scientific expeditions. In the early twenties, standard AM receivers were used, sometimes portables. Beginning in the mid- and late-1920s, however, newly developed short-wave equipment was taken along for two-way communication.

The seemingly frivolous displays of portable radios in the early 1920s—on police motorcycles, in boats and airplanes, in cars—played an important role in calling the attention of radio manufacturers to potentially lucrative markets for specialized portable or *mobile* radios. This era of experimentation in the use of portable radios helped launch the

7.4 Commercial portables of the late twenties. *a,* Kemper K-53, 1927 (16-5/8″ wide); *b,* Trav-Ler T-4344, 1929 (12-3/4″ wide); *c,* Automatic Tom Thumb portable, 1929

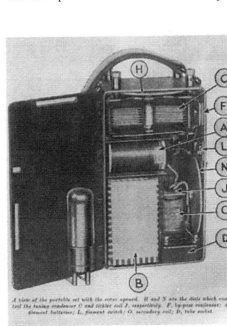

7.5 J. B. Armstrong's home-brew portable in a camera-style case, 1928. *Above,* man holding the assembled set; *right,* interior view

development of communications equipment that rapidly became essential to the functioning of various commercial, industrial, and governmental enterprises. Unlike the portables intended for the consumer market, the translation of images to mobile hardware often required costly and lengthy episodes of research and development.

Before radios—much less portable radios—could come to be regarded as a necessity by most families, Dxers would have to give way to listeners, a development that depended on progress in programming. Today we associate radio broadcasting with professional disc jockeys, newscasters, sportscasters, regularly scheduled programs, and incessant advertising. If we are old enough to remember the 1930s and 1940s, then radio also calls to mind afternoon soap operas as well as evening dramas, variety and quiz shows, and hilarious comedies. These features of radio's golden years developed gradually from the late twenties to the mid-thirties as the new medium matured.

In order for radio to achieve its full potential, the legal framework had to be modernized. The federal government, bound by the outdated Radio Act of 1912, was obliged to issue licenses to all comers. As a result, the airwaves in the early and middle twenties were chaos. Not infrequently, new stations in a city were assigned the same frequencies as existing stations. Imagine the thrill of hearing two operas at once—or twenty-three, the number of stations in Los Angeles on the same frequency.

A common solution to this problem was time sharing. For example, in the New York area stations WJZ and WOR broadcast on alternate days. In some cities, stations had only a few hours of time a week. Several clever operators simply took it upon themselves to transmit on a different frequency.

Secretary of Commerce Herbert Hoover, who was empowered by the Radio Act to issue licenses but little else, gradually exercised more control over the allocation of frequencies. Even this modicum of order was resisted. In a 1926 court case brought by Zenith, it was decided that the Commerce Department lacked a legal basis for Hoover's actions in regulating radio. A bad situation immediately worsened as "wave jumpers and pirates" colonized the ether. The patience of listeners was being tried mightily.

Congress finally acted in early 1927, passing the Radio Control Bill. It established a Federal Radio Commission (which later became the FCC) and gave the Secretary of Commerce an administrative role. His powers were far-reaching, and included the ability to refuse and revoke licenses as well as to assign frequencies, level of power, and hours of operation. Among Hoover's actions under this act was the weeding out of stations that he believed were not benefiting the public interest sufficiently. Not surprisingly, stations owned by powerful businessmen usually survived and were given the best frequencies and highest power ratings. Although he defined public interest narrowly to coincide with the interests of big business, Hoover did create order in the ether. As a result, the surviving broadcasters could turn their attention to programming and making radio pay.

Popular music underwent far-reaching changes during the twenties, partly in response to radio. In the early twenties, popular music meant songs from musical comedies, band music, sentimental ballads, and sopranos singing to a piano accompaniment. These varieties of middle-class music filled the airwaves along with high-brow opera and classical offerings. Younger Americans in the twenties were interested in hearing on the radio the same dance music that was available on phonograph records, but the most provocative tunes were kept off the air by a self-imposed censorship.

Jazz, of course, was at first totally taboo, for it was a musical form developed by southern blacks. Erik Barnouw, writing in *A Tower in Babel,* describes the early radio reactions to the new music: "Feelings against jazz seemed often to have an almost pathological dimension. William W. Hinshaw, preparing the WJZ audience for Mozart's *Impresario* went out of his way to denounce jazz as 'unhealthy' and 'immoral' . . . some stations forbade saxophones, which were assumed to have an immoral influence." Hostility to jazz quickly turned to acceptance as white musicians began experimenting with the possibilities of this amazing music.

If one person is to be singled out for contributing to the mainstreaming of jazz, it is Paul Whiteman. He brought respectability to jazz by wedding it to classical forms. It was he who commissioned George Gershwin to write "Rhapsody in Blue," a serious jazz composition that the Paul Whiteman orchestra first performed in 1924. By the end of the decade, a slightly sanitized jazz had been incorporated into the repertoire of the new radio bands led by Guy Lombardo, Ozzie Nelson, Rudy Vallee, and others. Jazz was now everyone's music.

A new style of singing called "crooning" arose in radio. Although Bing Crosby became its greatest practitioner in the thirties, crooning was originated in 1921 by a woman, Vaughn De Leath (who, incidentally, had begun broadcasting in the teens with Lee de Forest). The soft and gentle texture of crooning was an adaptation to radio technology of that day: by avoiding loud bursts in volume, the singer was less likely to blow tubes or bend the pointer on a meter.

Although changes in popular music helped transform radio's audience from Dxers into listeners, the crucial impetus was the rise of network radio. The practice of linking up stations to provide, temporarily, a single program in many states was not new. President Harding gave an Armistice Day speech in 1921 that was broadcast on an ad hoc network of stations linked by telephone lines. In 1924 the Democratic Convention in New York was heard over more than a dozen stations. For sixteen days anxious and weary listeners endured 103 ballots until, finally, John W. Davis was nominated. Who?

Two partial national networks were taking shape in 1925: one organized by AT&T and anchored by its New York station, WEAF, the other controlled by RCA and its anchor station, WJZ. Through a series of delicate negotiations orchestrated by Sarnoff, RCA acquired AT&T's network and agreed to lease telephone lines for the hookups. The National Broadcasting Company, owned by RCA, was set up to run the two networks (called "Blue" and "Red"). NBC began its regular broadcast schedule in January of 1927, reluctantly adopted advertising as a source of revenue (which Sarnoff had opposed), added stations on the west coast, and quickly became a success.

A competing network, founded in 1927, was called the Columbia Phonograph Broadcasting System. It floundered badly for lack of capital until it was rescued by Philadelphia cigar magnates William S. Paley and his father. The younger Paley at age twenty-six became president of the new network on September 26, 1928. Paley had exceptional managerial and financial skills, and within a year the network was prospering.

Vigorous competition between NBC and CBS (the phonograph was dropped from the name), was responsible for much creative programming in the late twenties and thirties. New talent, especially suited to the radio medium, came to the fore. For example, radio launched the careers of Bing Crosby, Kate Smith, and the Mills Brothers, all of whom became superstars on the basis of their *radio* performances. Rudy Vallee made his network debut in 1930, inaugurating a new genre of program—the variety show.

The emergence of radio stars and new kinds of shows brought radio to the verge of becoming a household necessity. The program that clinched it was "Amos 'n' Andy." "Amos 'n' Andy," we shall soon see, left a legacy far larger than the Kingfish's expression, "Holy Mackerel." The appeal of "Amos 'n' Andy" did have a dark side: it reinforced white stereotypes of "colored" people, keeping the latter at a comfortable social distance.

"Amos 'n' Andy" began as "Sam 'n' Henry" in January of 1926, on WGN in Chicago. A program that aired six days a week, "Sam 'n' Henry" followed the adventures of two southern blacks who had moved to Chicago to seek their fortunes. The show was uproariously funny; listeners laughed with characters trying to cope with a complex urban world mostly beyond their comprehension and control. Though white, many of the show's listeners—also immigrants and children of immigrants—could readily identify with the situations in which Sam and Henry found themselves.

The show was the creation of two white men, Freeman F. Gosden and Charles J. Correll, who not only wrote every word but also performed every role. They moved the program to WMAQ in 1928, leaving behind the old name (which WGN owned) and adopting the more mellifluous "Amos 'n' Andy." It became the first syndicated program in radio and was distributed on two twelve-inch records. At 78 RPM, it took both sides of both records to accommodate the fifteen-minute show.

"Amos 'n' Andy" first aired on the NBC Blue network on August 19, 1929, where it became an instant and spectacular hit. At 7:00 p.m., seemingly every radio in the country was tuned to the nearest NBC Blue affiliate. It is estimated that in 1931–32 "Amos 'n' Andy" was heard by forty million Americans; no other show so thoroughly dominated the ether. George H. Douglas describes this show's hold on an America plunged into depression: "In the summer months, almost anyone strolling down the streets of innumerable American towns could hear the sounds of the program wafting from window to window as he ambled along, and it might have been possible for this pedestrian to listen to the entire episode without going indoors."

"Amos 'n' Andy" attracted sponsors with ease, showing once-and-for-all that radio could be supported through commercials. In fact, by 1930, advertisers on the NBS Blue Network were paying $3,350 to sponsor an hour of prime-time programming.

Another commercial draw was sports, which became increasingly popular during the twenties. Not only did sports become an enduring feature of radio fare but it made enthusiasts of millions more Americans. Frank Conrad himself had read baseball scores during his home-brew broadcasts, and one of Sarnoff's early triumphs was to arrange for the broadcast of the Dempsey-Carpentier heavyweight fight. Appropriately enough, KDKA on August 5, 1921, carried the first live professional baseball game (the Pirates beat the Phillies, 8 to 5).

It is likely that the advent of radio "sportscasting" in the middle twenties helped to propel baseball and football into a place of prominence in American life. There is a rather impressive difference between reading about a game in the newspaper and listening, spellbound, to an announcer's animated play-by-play. The best sportscasters had, according to George H. Douglas, "learned the art of painting pictures with words, of putting the living and breathing essence of the sporting contest into the listeners' homes with verve and enthusiasm."

In the depths of the Depression still more kinds of programs, such as soap operas, quiz shows, and nightly news, were added to complete the mix of radio programming. Although Dxers continued to probe the ether for distant contacts, they were soon in the minority of listeners. Radio was now a reliable source of entertainment that could be enjoyed hour after hour without changing stations. In the Depression radio would become an audio hearth, a source of warmth and comfort to American families. In the evening everyone would gather to hear (and watch) the radio. The network programs had a broad appeal that drew in people—young and old, urban and rural, black and white, rich and poor. The visions of Bellamy and Sarnoff were fast becoming reality.

8

A Portable Depression

THE FALL OF THE STOCK MARKET IN 1929 WAS A NOT SO subtle signal that prosperity's bubble had burst. The next few years brought an economic decline of unprecedented proportions, which slowed but did not stop the penetration of new consumer goods into American homes.

The Great Depression began as a panic of capitalists, many of whom took a beating in the stock market crash or had friends who did. In 1929 business investments in the economy reached $16.2 billion, but only two years later it was $800 million—a drop of 88 percent. When the economy bottomed out in 1933, construction was down 78 percent compared to 1929, while business investments declined a whopping 98 percent.

The capitalist panic soon began to affect everyone else. Though not wiped out by the crash, most ordinary Americans became cautious consumers, saving (or hoarding) rather than spending. In 1930 and 1931, more than 3,000 banks failed, further eroding consumer confidence. By 1931, deflation set in: demand, production, wages, jobs, and even prices fell simultaneously. Ironically, even though people had less money to spend, it bought more. Food prices declined sharply in the early thirties, making it possible for those not dependent on farm incomes to sustain the culture of consumption. The most devastating statistic was unemployment, which went from 3.2 percent in 1929 to a staggering 24.9 percent in 1933. One out of every four working Americans was jobless. But mere numbers do not really reveal the extent of human suffering. As the search for work went from weeks to months and from months to years, able-bodied people lost their sense of self-worth. Many became homeless, living in squatters' towns dubbed "Hoovervilles."

Herbert Hoover (another Republican) was elected president in 1928 after Coolidge refused to run. Because the Republicans, in office throughout the twenties, had taken credit for the prosperity of that decade, it was only fitting that their man in the White House also accept blame for the economic collapse. In the elections of 1932, Hoover and the Republicans were tossed out with a vengeance. The New Deal had begun.

Franklin Delano Roosevelt was a wealthy member of the New York landed gentry who, because of his bout with polio, had a genuine empathy for the suffering masses. Despite his patrician background, he was *loved* by poor Americans; his picture, along with that of Jesus, occupied a place of honor in many homes. Roosevelt also had an immense personal charm, which contributed to his popularity. No stranger to radio, for he had given a nomination speech for Alfred Smith (candidate for governor of New York) in 1926, Roosevelt took almost immediately to the airwaves.

Shortly after his inauguration on March 4, 1933, Roosevelt began his "fireside chats." He understood, perhaps better than anyone, the role radio played in politics; he appreciated that a special style was needed for addressing people on intimate terms in their own homes, usually beginning each broadcast with "My friends. . . ." Roosevelt's radio style was enormously effective, as Robert S. McElvaine recounts in *The Great Depression*: "Even lifelong Republicans and veteran Roosevelt-haters have testified that while listening to his broadcasts they sometimes weakened to the point of almost believing him—a condition from which they usually recovered by the next morning when they read newspaper accounts of what he had said." His reassurances that the banks were sound had a positive effect. Deposits and even the stock market went up.

In 1933, as the New Deal started in the depths of the Depression, sixteen million families—almost half of all Americans—could hear the President on their radios. When Roosevelt earnestly told his millions of friends that "the only thing we have to fear is fear itself," he had touched a sizable proportion of the electorate.

Acting quickly, Roosevelt and the democratic Congress passed a flurry of legislation aimed at restoring confidence and providing relief for the jobless. The Public Works Administration undertook large construction projects, putting millions to work on bridges and dams; this agency also erected 70 percent of all new educational buildings in the United States between 1933 and 1939. Similar programs employed millions more, though wages were not tremendous. In the mid-thirties, a subsistence level for a family was deemed to be $1,200 a year; annual relief wages ranged between $624 and $1,300. Clearly, these programs did not foster a sumptuous lifestyle, but they restored dignity to the down-and-out and, by putting money in peoples' pockets, boosted demand for commodities. By 1937, production finally exceeded 1929 levels, though millions were still out of work. Despite its good words and good deeds, the New Deal did not end the Great Depression. Complete relief would not be achieved until 1941, when the United States entered another world war.

In the meantime, to spur spending, manufacturers kept the new industrial designers—Raymond Loewy, Norman Bel Geddes, Walter Dorwin Teague, and so on—working overtime to remake their products so that they looked modern; that is, appeared to be in harmony with the efficiency of the machine age. The distinctive style that they forged in the early thirties was streamlining, obviously influenced by aerodynamics. Writing recently, Arthur J. Pulos made clear the appeal of the new style: "Streamlining was the first new and uniquely American approach to form that the public could associate with progress and a better life."

Over the decade, streamlining was applied relentlessly to virtually everything, including gas stations, locomotives, refrigerators, bar stools, typewriters, pencil sharpeners, and cigarette lighters. Needless to say, radios were also streamlined. For the first time, the products of American industry shared a unifying style. Streamlining made a contribution to fostering demand for many things; however, even without the new style, radios would have continued to enter American homes in large numbers during the Depression.

The coming of network radio, the all-electrics with one-knob tuning, and the greater stability of stations meant one thing to the average American by the beginning of the new decade: radio had become a necessity. This belief was dramatically underscored when Roosevelt began his fireside chats. To be without a radio was to be without entertainment, economic news, and, more importantly, a personal link to the president. However, as noted in Chapter 7, the all-electrics were very expensive, and clearly beyond the reach of many families. Predictably, radio sales plummeted with the onset of hard times. Retail sales of radios dropped to around $200 million in 1930; moreover, radio makers had overproduced, and elegant consoles (Fig. 8.1) were a glut on the market.

The worsening economic situation did, however, create opportunities for entrepreneurs on the fringe, outside the large, well-established radio companies. In Los Angeles, many such firms began to sell smaller, less expensive sets—the so-called "midgets"—better suited to tough times. In the East it was Emerson, a company that prospered during the Depression making down-sized radios. (The term "midget" was actually applied to a range of radios, from quite small table models to majestic cathedrals and "tombstones," the majority of which—with built-in speakers—were small compared to the all-electrics of the late twenties. It is these wooden tombstones and cathedrals that today so powerfully evoke nostalgia for that bygone era of classic radio.)

The small companies that seemingly came out of nowhere in 1930 began carving out a significant slice of the radio market. With their growing backlog of unsold consoles, members of the radio establishment sat on the sidelines, stunned by the midgets' sudden success. By the late thirties, the surviving majors all were making midgets, and a few of the upstart midget makers were among the majors. How did these companies score such a victory in the marketplace?

At first the midget makers achieved economies by reducing the number of components to the absolute minimum, cleverly arranging the components on a smaller chassis, and shrinking the speaker and cabinet. In the first generation of midgets, from 1930, prices ranged from $50 to $90 (these sets had from five to eight tubes).

When it became clear that midgets were going to remain an important genre of home radios, manufacturers invested profits in research, which led to useful electronic innovations. Indeed, beginning in 1932–33, midget makers approached the design of AC sets from a new perspective. Instead of producing a bulky radio with a built-in battery eliminator (which required a sizable and expensive power transformer), they designed sets that took their power directly from the line. Filaments of new

8.1 Unsold consoles like this General Electric Highboy were a glut on the market in 1930

8.2 An array of 1933 RCA midgets

tubes were connected in series and the excess voltage consumed with a resistor; the line also supplied B voltage, which was converted to DC with a rectifier tube. Dropping the power transformer reduced the radio's size, weight, and, above all, price.

Working with suppliers, the midget makers also managed to shrink other components. Prior to this time, there had been no effort to reduce the size of volume controls, tuning capacitors, intermediate frequency transformers, and so forth. But once it became a priority, miniaturization was readily accomplished; what is more, the smaller components used less raw material, and so were cheaper. In this way, the entrepreneurial fringe was able to spur electronic innovation and produce a decent, inexpensive radio. In 1933, the price of midgets descended to between $20 and $50. Remarkably, a few set makers, including Crosley and RCA had four- and five-tube superheterodynes for just under $20 (Fig. 8.2). In the late thirties, some sets crashed through the $10 barrier. Though obviously not putting out the powerful, mellow tones of a console, the midget was a fine second set and, more importantly, the perfect product for the poor family that had no other radio. Coming out of a midget, the jokes of "Amos 'n' Andy" were just as funny.

The second factor that contributed to the success of the midget makers was their easy access to retailers. Radios were sold to the public by both family-owned radio stores and large retail chain stores like Sears, Montgomery Ward, and Western Auto. A new radio company could reach the small dealer directly by advertising in trade magazines such as *Radio Retailing*. Owners of radio stores were always looking for new products that promised higher profits, and buying straight from the manufacturer cut costs. (Most of the majors relied on wholesale distributors to market their wares to franchised dealers.) Mass merchandisers like Sears were equally receptive to overtures by new manufacturers because they made no radios themselves. Production of their "house brands" (like Montgomery Ward's Airline and Sears' Silvertone) was farmed out, usually to the lowest bidder among interested set makers.

Look at this Value! Model R-28, RCA Victor "Carryette." One of our series of the finest miniature radios ever made. Five tubes, tone control, electro-dynamic speaker, also police waves, beautiful cabinet. Only **$19⁹⁵**

A Personal Radio for Your Bedroom! RCA Victor "Carryette" Model R-28A. Also with 5 tubes, tone control, police waves, superhet. Walnut finish. Only **$21⁵⁰**

Another Splendid Performer! RCA Victor "Carryette" Model R-28B. 5 tubes for tone, power, performance, plus police wave. Lovely walnut-veneered Personal Box. Only . .   **$24⁹⁵**

Thus, new companies with a good product could, albeit anonymously, gain some market share. (That the retail system was so open to outsiders proved helpful decades later to Japanese companies when they began selling portable radios in the United States—see Chapter 14).

Despite the severe depression, then, midgets made it possible for more families to buy radios. It is estimated that by 1937 almost twenty-three million American families had a radio—far more than had electric refrigerators, telephones, electric irons, and vacuum cleaners. Even the cheapest midget of the early thirties was easier to use and performed better than most sets of the *early* twenties. Radio was now positioned, price-wise, to enter all American homes, and this it did by the end of the decade (when midgets were called "table models").

Radio had clearly become indispensable to a family's emotional and social well being. People who lived through the Depression speak glowingly about the importance of radio, the spell it cast as it brought fantasy and humor and tragedy to the living room. And how fondly they remember their family's first radio, perhaps bought with scrupulously marshalled savings or received as a hand-me-down from a friend or relative. No matter, for by this time almost everyone lived within range of a powerful network station, and so even the most feeble midget or twenties battery set could capture Burns and Allen, Jack Benny, Edgar Bergen and Charlie McCarthy (Fig. 8.3), and the other comedians.

During the Depression David Sarnoff's lofty vision of radio's role was realized as the national networks knit Americans into a unified nation. Despite differences in dialect, food, and lifestyles, which varied regionally, and differences in wealth and social position (present in every community), Americans came to share certain values and outlooks promoted in radio shows. Radio created a national consciousness as never before.

Radio had joined magazines, books, and movies as a medium of mass culture. But those older forms of entertainment did not wither and die under radio's impact. During the Depression, they also thrived. Through these media, old cultural imperatives were updated, new visions of the future were advanced, and the central tenet of the consumer culture— that new products would create the good life for all families—was endlessly reinforced. With sound, movies could captivate an audience as never before, and that audience grew yearly. By 1937, theaters were drawing an audience equal to the entire U.S. population every ten days. Movie-goers watched musical extravaganzas as well as comedies and dark dramas. Science fiction of every kind flourished in Depression movies. *Frankenstein* (with Boris Karloff as the monster, 1931), *King Kong* (1933), *Dracula* (with Bela Lugosi as the count, 1931), and *The Werewolf of London* (1934) were among the celluloid fare that provided a scare—as if the plummeting economy were not enough. Both horror movies and the economy gave Americans the same message: mysterious forces beyond the control of ordinary people can do great harm.

Buck Rogers, whose first installment had appeared in Gernsback's *Amazing Stories* in 1928, also took to the silver screen in the early thirties. Among the paraphernalia of this twenty-fifth–century hero was the walkie-talkie—a kind of portable radio. In 1932, Buck's exciting adven-

8.3 Edgar Bergen and Charlie McCarthy in David Sarnoff's office (RCA publicity photo)

tures entered American homes in fifteen-minute radio episodes, three times a week.

In popular literature, many now-familiar characters first appeared during the Depression. Flash Gordon, another high-tech hero of the future, debuted in the comics in 1934. In short order he too would be seen by millions in the movies. Superman comics began in 1938; this superhero of the present embodied a new, more optimistic mood. Isaac Asimov and Robert Heinlein started publishing stories in 1939, which launched the modern era of science fiction. Both men had scientific training, and their writings vigorously promoted the belief that science and technology would bring a wondrous, gadget-filled future. The New York World's Fair of 1939 fixed all eyes on the future, and pushed that same vision.

While the public consumed a popular culture that promised better times—and better things—ahead, scientists and engineers were busily perfecting some of the products long envisioned by the techno-mancers. It has often been observed that the 1930s was no depression for science, technology, and radio in particular. Despite declining sales at the beginning of the decade, the radio industry was optimistic. Both RCA and General Electric invested heavily in research during the 1930s, and a spate of patents and new devices flowed from their laboratories. Even in some of the smaller companies, often poorly capitalized, engineers were trying to find that one gadget or gimmick that could create enormous wealth.

Television was the *Wunderkind* of the decade, which finally matured just before the war, both in England and the United States. Many people and companies made substantial contributions to the development of U.S. television, including Philo Farnsworth, Allen B. Du Mont, and Philco, but RCA (with David Sarnoff as its president) usually takes—and is often given—the lion's share of credit. In early 1923, David Sarnoff had dashed off another bold memo, this time to RCA's Board of Directors. In it he outlined his belief that television "will come to pass in due course." The Orpheus of RCA continued:

> Radio development will provide a situation whereby we shall be able actually to see as well as read in New York, within an hour or so, the event taking place in London, Buenos Aires, or Tokyo.
>
> I also believe that transmission and reception of motion pictures by radio will be worked out within the next decade. This would result in important events or interesting dramatic presentations being literally broadcast by radio through the use of appropriate transmitters and, thereafter, received in individual homes or auditoriums, where the original scenes will be re-enacted on a screen with much the appearance of present-day motion pictures.

Actually, the idea of television itself was not new—far from it. Television, given its English name in 1909 by Hugo Gernsback in *Modern Electrics*, was rife in novels of the 1880s and 1890s, including mentions by Jules Verne in 1885, Camille Flammarion in 1893, and H. G. Wells in 1899. Through fiction and the imaginative prophecies of technically trained people, television had become a cultural imperative by the end of

the nineteenth century. A great many people fervently believed that this invention would come to pass and bring enormous benefits, as had the telegraph and telephone. Tele*vision*, after all, was a logical extension of these communication technologies.

Clever tinkerers in Europe and the United States had begun to design and build mechanical systems of image pick-up and projection, as early as Paul Nipkow in 1884, and some successes were achieved. A limited number of TVs had actually been marketed in the twenties. In 1928 Hugo Gernsback, feverishly trying to bring his own prophecies into being, began daily television broadcasts from his New York radio station, WRNY.

Televisions of the twenties had a small, flickering image that seemed to be a permanent liability of the partly mechanical systems (Fig. 8.4). Sarnoff, as usual, had been the one to recognize that radio engineers would soon be able to build a high-quality, *all-electronic* system. In early 1929, Sarnoff put the RCA research facilities and staff at the disposal of Vladimir Zworykin, a Westinghouse engineer who claimed that he could develop all-electronic television. Later that year Zworykin exhibited his iconoscope, the first television camera tube. When asked by Sarnoff how much it would cost to create a practical television *system*, Zworykin confidently replied $100,000. Zworykin was put to work.

In the next few years, problems in developing *practical* all-electronic television turned out to be more difficult and costly to solve. And new problems kept arising. By the end of the thirties, when success was finally at hand, RCA had already spent almost $10 million. This company would ultimately invest more than $50 million before television started to turn a profit.

Throughout the 1930s, television was literally "in the air" through experimental broadcasts. By 1938, when there were seventeen experimental TV stations, some regular programming had begun. RCA launched its television system with much fanfare in 1939 at the New York World's Fair, and televised Roosevelt's speech opening the Fair. The RCA exhibit building (in the shape of a giant vacuum tube) housed thirteen television receivers, including a six-by-ten-foot projection set. The theme of the fair was "The World of Tomorrow," which gave America's industrial giants the opportunity to wow visitors with products of the future (like television). Admission to the Fair—to the future—cost fifty cents.

Regular television broadcasts were begun from RCA's transmitting antenna high atop the Empire State Building, then at ninety-eight stories, the world's tallest. Prospective viewers could buy an RCA receiver for $615, and hundreds did (Fig. 8.5). (Du Mont, however, was first to market an all-electronic TV in the United States.) It is estimated that in New York City 10,000 people watched the excitement of the 1940 Republican National Convention from Philadelphia, which nominated Willkie on the sixth ballot. Television was an impressive technological success. The Second World War, however, delayed its success in the marketplace.

It is difficult to imagine the president of a U.S. corporation today investing in a product that would not produce profit for more than two decades. Sarnoff's strategy in running RCA was to put all profits into expansion and product development. Though a very prosperous company, RCA paid no dividends until 1937. Stockholders had to be content

8.4 Ernst Alexanderson (*left*) and assistant watching an early RCA mechanical television, 1928 (RCA publicity photo)

with their appreciating equity. The financial markets in the U.S. today, where the half-life of an investment seems to have declined to about ten minutes, make Sarnoff's long-term strategies seem quaint. Carping from the board of directors, stockholders, and Wall Street analysts would dissuade even the most courageous visionary from engaging in such long-term folly. Yet, for RCA, Zenith, and other technology-based corporations, the long view worked extremely well. (Not surprisingly, Japanese corporations today plow back profits into decades-long development projects.)

During the Depression, television would have seemed fantastic to a poor farm family, still without electric lights, listening to a second-hand, battery radio. Americans such as these probably wondered who would be able to afford televisions when they were finally sold on a large scale. Manufacturers sometimes wondered about this too, for the all-electronic television, with dozens of tubes and hundreds of components, would not be cheap. World War II deferred this worry, contributing to the prosperity that within a decade brought TVs into most American living rooms.

Because television was imminent during the immediate pre-war years, manufacturers were concerned that prospective buyers, waiting for television, might put off buying new radios. In an ad in *Life* magazine in

1939, RCA reassured consumers that if they bought a new RCA receiver, it would not become obsolete, provision having been made for television: "You will simply connect a Television Attachment to [the new RCA radio shown in the ad]. Then the Attachment will reproduce the pictures . . . the radio the sound." One of RCA's first TVs did indeed plug into radios, some of which had jacks on the back labeled "television." However, most TVs were built as complete systems. And radio did suffer significant impacts from television, as we shall see in Chapter 13.

Though the thirties was an important era for visions and real electronic innovations, portable radios were an uncommon commodity. Only a handful were made between 1930 and 1939, and advertising for portables all but disappeared from national magazines. The Depression "portables," though few and far between, are an interesting lot because most were not true portables and because they foreshadowed design features that would become standard in 1939.

In the thirties appeared a small number of sets with handles, seemingly ready to travel; some even had panels to cover the controls. Although advertised as "portables," these radios could not be powered by batteries; I call them "pseudo-portables." Typical was the 1934 Majestic Knockabout, envisioned as an "all purpose" small radio that could be moved about easily from home to office and so forth (Fig. 8.6a). The Knockabout was encased entirely in leatherette with a handle and latching cover on the front; when closed, it gave no clue to its identity as a radio. The Knockabout, however, was not a battery portable but a highly portable midget.

Other examples of this genre came from Freed-Eisemann, which in 1936 and 1937 sold several pseudo-portables (Fig. 8.6b–d). The FE-26, for example, was called a "Traveling case portable." Like the Majestic Knockabout, this set was a midget, powered by 110 volts, that could travel with ease. The similarity ends there, though, for the Freed-Eisemann 26 had a distinctive appearance. The radio was "available in black or brown fabrikoid or blue, green, red, or brown striped airplane cloth." In effect, Freed-Eisemann had put their "portable" into a small, fashionable suitcase. The Espey 861B (Fig. 8.6e) also sported an airplane cloth exterior. Though not battery sets, the pseudo-portables of the thirties (especially those from companies on the entrepreneurial fringe) established design conventions that would be maintained when true portables re-emerged en masse at decade's end.

From the vantage point of today, it seems odd that these sets would have been called "portables," since they could not play on batteries. From the standpoint of the early 1930s, however, the portable label made a certain sense. When the all-electrics appeared in the late twenties, sets became uniformly heavy and thus much less mobile than the light—and often smaller—battery sets sold earlier in the decade. Relative to the behemoth all-electrics, then, small midgets with handles would have seemed eminently portable.

Another genre of Depression "portables" was known as "universals"; I prefer to call them "convertibles." These were operable on AC or DC (as in cars, boats, or on 32-volt farm systems) but had no provision for *inter-*

8.5   RCA's first all-electronic TV receiver, the TRK-12, in a 1939 publicity photo

a

b

c

d

e

8.6   Pseudo-portables of the Depression.
a, Majestic Knockabout, 1934; *b,* Freed-
Eisemann FE-26, 1937; *c,* Freed-Eisemann FE-
33, 1936; *d,* Freed-Eisemann FE-58, 1936; *e,*
Espey 861B, 1938

nal batteries. A few such models were sold every year from 1930 to 1936,
a time of great interest in add-on car radios. One of the most significant
convertibles was manufactured by International Kadette, of Ann Arbor,
Michigan, beginning in 1932 (Fig. 8.7, below). It was a 4-tube midget, the
first small radio with a Bakelite (plastic) cabinet to be produced in large
quantities. The Kadette could be powered by house current or by external battery; one ad showed the set attached under the dashboard of a car.
Because it lacked a handle and had no space for batteries inside, the
International Kadette was not a true portable. Nonetheless, these sets—
at $25 to $30—sold well, and encouraged the Bakelite Corporation to
push other radio makers hard to adopt plastic cabinets. Many midgets
had plastic cabinets by 1940.

In 1933 (the same year Gernsback began publishing *Sexology*) International Kadette brought out another convertible, one of the smallest made
with tubes—two of them, to be precise. It was called the Kadette, Jr.,
and had a Bakelite case. The ad in *Radio Retailing* claimed that it was
"the *first* and *only* pocket radio." To achieve this degree of miniaturization (it weighed only two pounds), even tinier components had to be
produced. *Popular Mechanics* ran a brief piece on the Kadette, Jr., acknowledging that it was a "pocket-size" radio (well, coat pocket). Signifi-

cantly, the article showed the set being used by children (Fig. 8.7, above). This still-life conveyed the impression that a set this small was only fit for small fry. Despite national advertising and a low price ($12.50), the radio sold poorly and was discontinued in 1934. Nonetheless, the Kadette, Jr., achieved appreciable gains in miniaturization and hinted that a true pocket portable was beginning to enter the realm of feasibility.

Another convertible was the General Electric B-52, advertised as a "Portable Auto Radio." It had a homely metal cabinet shaped like a stretched-out toaster with one end sloping like a snout; but the B-52 had, the ad claimed, "full, brilliant tone." Another convertible was the $49.50 Motorola "Companion"—a name that harked back to Zenith's first portable. Like the General Electric B-52, the set had a metal case and handle, though it lacked a snout.

Tracking down the true portables from the early and middle thirties is frustrating because most of the sets having handles and advertised as portables were midgets that worked only on house current or were convertibles operable on *external* batteries only. On the other hand, many battery-only sets turn out, upon close inspection, to be farm radios, not designed to be portable. For example, in 1938 General Electric offered a number of "battery radios," including a lovely cathedral and a stately console—neither with a handle. These battery sets were mainly targeted at the farmer, though General Electric also envisioned their use on "boats and yachts."

One definite portable was made by RCA in 1932, the P-31. It was an eight tuber in a large wooden cabinet with a handle. In 1934 Simplex Radio sold a "perambulator radio"—a baby carriage with a built-in receiver; it did not catch on. In 1936, there was a slight stir of interest in true portables. *Radio Retailing* of June contained a one-page article entitled "Practical Portables." It noted improvements in battery tubes and reductions in the size of dry batteries, which "make summer sets easy to carry, economical to operate." In the accompanying picture, a young couple was shown in a canoe, staring at—and presumably listening to—their new portable radio.

The set was a four-tube superheterodyne, the Ansley B-1, which cost $39.50. Like some of the pseudo-portables and convertibles, the Ansley B-1 had a fold-up front and an airplane cloth covering, and it resembled a small suitcase. Ads to radio dealers (Fig. 8.8a) pitched the portable as an "outdoor radio" that could be sold during the summer slump, when home radio sales predictably plummeted. Other true portables of 1936 are shown in Figure 8.8b and c.

The few true portables of 1936 were manufactured by companies on the entrepreneurial fringe, which did not advertise their sets in mass-circulation magazines. Nor did the latter take any interest in the new radios. True portables did not become a craze, though they were electronically sound and played reasonably well. Most telling of all, the major radio makers did not jump on the portable bandwagon, for it did not get rolling. In 1937 and 1938, true portables were again slumbering in oblivion. Only the far entrepreneurial fringe had any offerings, one of which—the Eastern RC-15—came and went quickly (Fig. 8.8a).

8.7 International Kadette Convertibles. *Above,* Junior, 1933; *below,* Midget, 1932

a

b

c

The true portables made between 1930 and 1938 have left little trace today. The genre was such an oddity during the Depression that when Pilot and Philco introduced their earliest models at the end of the decade, both companies claimed in ads to have developed the first portable battery sets. Why did the true portable languish in this period even though it was possible to make a respectable set that would have sold for under $50?

It is tempting to conclude that the Great Depression itself was responsible for the portable radio's demise. This was, after all, a time of belt tightening; the portable radio was hardly a necessity and would have been a purchase easily deferred. This explanation is comforting, but does not stand up well. Though poverty was widespread and countless families were uprooted, it is also true that many Americans still had lots of discretionary income.

Luxury products did not disappear during the Depression—far from it. Expensive cars, furs, jewelry, and so forth found eager buyers. Moreover, wealthy folks with high-tech interests had a choice of many new gadgets, including home-movie equipment, short-wave transmitters and receivers, electronic organs, expensive hi-fi sets, and, of course, a variety of experimental televisions that were obsolete almost immediately after purchase. Radios themselves were available with new—and expensive—gimmicks, including remote controls, electronic push-button tuning, and tuning eyes. In 1937 RCA offered its 9K3 console model equipped with no less than "Magic Voice, Magic Brain, Magic Eye, [and] metal tubes." If that set was too tame, Capehart had a model, the 500F, sure to

d

8.8 True portables of the Depression era. *a,* Eastern Radio and Television Company, 1938; *b,* Simplex Sportsman, 1936; *c,* Freed-Eisemann P-55; *d,* Ansley B-1, 1936

please elite tastes. It was a twenty-seven-tube radio-record player combination that could automatically play twenty records—on one or both sides. The huge wooden cabinet was as spectacular as the price: $2,500. Clearly, upscale purchasers could still find in the depressed economy a wide range of products befitting their social positions. For more than a small minority of Americans, then, the Depression was a temporary annoyance that would pass. In the meantime, life—particularly shopping for luxury goods—did go on.

Conceivably, then, fancy-looking portables could have been marketed as playthings for the rich. Radio makers, however, apparently did not believe that true portables would now find favor among the wealthy, despite the simpler operation and more compact design that was possible in the early thirties. Probably the memory of the unhappy fate of twenties' portables was still too fresh. After all, the same generation of company managers and engineers that had seen the earlier portables peter out still held power, and in those uncertain times was unwilling to chance another marketplace fiasco. That these executives may have been right was indicated by the limited success of the few true portables made in the thirties. Had those sets sold well, major radio makers would have added true portables to their lines, just as they had earlier added midgets. Even RCA, with Sarnoff as president, brought out only one true portable between 1926 and 1939.

Clever people in the twenties had shown that portable radios did have important roles to play in industry, commerce, public safety, and so forth. During the thirties, specialized "mobile" radios made a great impact as they entered police cars, fire trucks, taxi cabs, boats, and even airplanes.

One interesting British set of this time was designed for policemen on the beat. Very compact, it measured only 6-1/2″ × 4-3/4″ × 2″ and weighed less than two pounds—with batteries. The antenna, however, was not built in but had to be worn inside the back of the officer's coat. When this miniature radio was reported to Americans (in *Electronics*), the article suggested that there might be a market in the United States for the "pocket portable radio set." Radio makers took no notice.

The auto radio, too, became a familiar fixture in the thirties. In 1935 more than one million add-on car radios were sold by forty-two companies at $30 to $70 each (Fig. 8.9). By the end of the decade, 20 percent of cars rolling off Detroit's assembly lines had built-in radios. In these specialized guises, the portable radio thrived. But these sets, installed permanently, ceased quickly to be portable; mobile yes, portable no.

Another use of portables established in the twenties was as a radio that could be easily moved from home to office and back. This function was readily served by the midgets and pseudo-portables, which were easy to carry. There was simply no need for such "portables" to run on batteries. Similarly, the development of short-wave sets by Grebe, Zenith, and other companies in the 1920s met the needs of exploring expeditions.

It would seem that the commercial functions established for portable radios in the early twenties were, a decade later, being served by specialized mobile and short-wave equipment. Still other functions (the capability of being moved between locations that had power) were car-

"His Master's Voice" on the Road!

THIS IS ALL THERE IS TO IT!
Only $39.95 Complete
HAVE IT INSTALLED WHILE YOU WAIT

8.9   RCA add-on car radio, 1933

ried out by midgets with handles. New electronic gizmos catered to the social needs of the trendsetters. Radio companies apparently believed there was simply no real radio work left for the true portable; new uses would be required before it could be revived. In effect, the portable killed itself in the late twenties by calling attention to markets that could be better served by more specialized battery radios.

As a cultural imperative, however, the portable radio survived—and not just in fiction. Another generation of young experimenters was building their own portables during the thirties. Their unique sets—like those of the teens and twenties—perpetuated the belief that small, battery-powered radios, easily carried about, would someday become a casual companion to people at work and play.

Hugo Gernsback was editing a new magazine for electronics buffs in the thirties, *Radio-Craft*, having lost control of *Radio News*. (Always ahead of his time, Gernsback went bankrupt the year before the stock-market crash.) In *Radio-Craft* Gernsback continued to push portables to youthful enthusiasts. In the May issue of 1933, Ulysses Fips, Staff Reporter (one of Gernsback's pseudonyms), described "A Revolutionary Radio Development—the Vest-pocket 7-Tube Superhetero-Ultradyne." The mini-radio could be held in the hand and incorporated new mini-tubes about an inch long and 3/8 inch in diameter. It was called a WestingMouse. The irrepressible Gernsback had played a joke on his readers (the tubes were numbered APR-1). As usual, though, he got the last laugh at the end of the decade when tubes that small were actually made commercially (see Chapter 12).

Other designs for home-brew portables were more immediately feasible. For example, in response to many reader requests for a compact portable radio, *Popular Mechanics* furnished a four-tube design in 1934 (Fig. 8.10d). The article modestly claimed: "exceptionally easy on batteries, simple to build and the parts are not expensive. Distant stations come in with good loud-speaker volume and the tone is quite satisfactory for a portable receiver."

a

TO BATTERY BOX

TUNING CONTROL

VOLUME CONTROL

b

c

8.10 Do-it-yourself portables of the thirties. *a,* bicycle radio, 1936; *b,* Camera Radio, 1933; *c,* A. J. Haynes' "Pocket Sportset," 1937; *d, Popular Mechanics'* beach set, 1934

Popular Science Monthly presented plans in 1936 for a one-tube "Wrist-Watch Radio." Not only was the radio rather large, but the wearer needed a belt to carry the batteries. Two years later the same magazine featured a vest pocket "Fountain-Pen Receiver," which employed one British hearing-aid tube (and also had external batteries). Camera-style radios (Fig. 8.10b) and even bicycle radios (Fig. 8.10a) also made appearances in hobbyist magazines.

A. J. Haynes in 1937 showed the readers of *Radio News* how to build a one-tube "pocket sportset," acknowledging that "the idea of a pocket receiver is not new." The headphone-only set—displayed on the magazine cover—was built in a tiny plywood box; small enough, it was claimed, to slip into a coat pocket (Fig. 8.10c). The main use of the pocket portable was envisioned as follows:

> Haven't you often wished, when hearing an expert describe a World Series baseball game, a football match, or a prize-fight over the air, that you could listen to that same vivid description while actually sitting in the stands watching the action? . . . You can do just this with the Pocket Sportset.

Clearly, electronic hobbyists were already thinking up—*and trying out*—new uses for portable radios. In the decades ahead, such uses would help to create a place for the portable—even pocket portables—in the lives of ordinary Americans.

Though true portables would not be common until 1939, other modern consumer products became important during the Depression, and still others were being forecast by techno-mancers. In the late thirties, Scotch tape, fluorescent lights, nylon stockings, and other near-miraculous products went from corporate laboratories to store shelves. Sales of electrical appliances made steady headway during the Depression, but none yet achieved the popularity of radio. By the end of 1936, American roads supported more than twenty-four million passenger cars—nearly one per family.

During the thirties airships and airplanes provided expanded service for the well-heeled traveler. In 1936, scheduled airlines carried 100,000 passengers per month, a startling increase over the previous decade, when there were only 5,800 passengers *all year*. As Gernsback had predicted, airplanes were clearly eclipsing airships for passenger travel, and the latter would soon disappear. On May 6, 1937, the enormous airship Hindenberg, while attempting a landing during bad weather at Lakehurst, New Jersey, exploded and burned. Among those on hand to greet the arrival of the Hindenburg was a radio reporter, Herbert Morrison, from WLS in Chicago. The following excerpts indicate why his live broadcast describing the death and destruction became one of the most heart-rending news accounts ever heard on radio:

> The ship is riding majestically toward us, like some great feather, riding as though it was mighty proud of the place it's playing in the world's aviation. . . .

> It's practically standing still now. . . . The vast motors of the ship are just holding it, just enough to keep it from. . . . It burst into flame. . .

> It's crashing, terrible . . . it's burning, bursting into flames and it's falling on the mooring mast and all the folks between us. Oh, this is terrible, this is one of the worst catastrophes in the world.

Many people died in the fiery debris, and so did airship passenger service. The Hindenburg accident was simply the last nail in the airship's coffin. For reasons specified by Gernsback a decade earlier, the airship was already an antique.

One of the most thoughtful techno-mancers of the Depression was S. C. Gilfillan, among whose 1937 prognostications were color television, 3-D movies, and "the early arrival, probably within 5 or 10 years," of stereo in radios, phonos, and movies. His predictions for cable television turned out to be uncannily accurate. Cable TV, he said, "will broadcast all sorts of visible entertainment not sufficiently popular to win the very scarce broadcasting channels, but which could pay for themselves through charges for the wire service. . . . Censorship will be less important with this wired television than with radio."

Other techno-mancers forecast FAX in the home, "smellevision" (television sets equipped to emit appropriate odors for each scene), and luminescent paints that store sunlight during the day and glow at night. These new products would surely brighten the future.

9

The Portable Radio Revival

AS THE NEW DEAL BEGAN TO EASE THE EFFECTS OF THE
Great Depression, disturbing news from Europe and around the world
arrived at American homes on invisible waves. The rise of Nazism under
Hitler, with its overt territorial ambitions, could not have brought comfort to a generation of Americans, now entering middle age, who had
already fought in one European war against Germany.

Although World War I ended before the advent of commercial broadcasting and the modern radio age, its lessons were widely learned. The
"war to end all wars" was the first foreign engagement that involved a
sizable number of Americans; sons and grandsons and nephews went off
to strange lands, and many returned home wounded or in pine boxes.
This war demonstrated above all that distant events, not controlled by
the United States, could have a drastic effect on one's own family. And
that lesson was indelibly etched in the minds of Americans, especially the
young men returning in victory.

In seeking re-election in 1936, President Roosevelt had pledged to
keep America at peace. However, in October of 1937, after being re-
elected, Roosevelt decried "the present reign of terror and international
lawlessness," adding ominously, "Let no one imagine that America will
escape, that it may expect mercy, that this Western hemisphere will not
be attacked." Hearing these words over radio, the veterans of World War
I, many of whom now had sons of their own approaching draft age, must
have trembled with that special anxiety only parents can feel. Nonetheless, the dominant mood of the country was still isolationist.

With the coming of the high-circulation national news magazines—
Time, Newsweek, and *Life*—and especially radio, America had become
a nation of news junkies. This did not happen all at once, of course, but
it was inevitable after the mid-thirties when radio networks began to
bring world news—sometimes live—into the home. Foreign correspondents, such as John Gunther in London for NBC, William Shriver in
Berlin for CBS, and Edward R. Murrow, also for CBS (in London), became
as familiar to Americans as Clark Gable and Babe Ruth. By the end of
the 1930s, each new aggression by Japan, Italy, and Nazi Germany was

being meticulously recounted and interpreted to a large and attentive radio audience.

The pace of major events quickened as Germany annexed Austria in 1938. In September, leaders of the major European powers met with Hitler at Munich. In one of the most infamous diplomatic episodes in history, Germany gained great concessions, including the Sudetenland— a large chunk of Czechoslovakia—through a policy of appeasement which, it was hoped, would prevent wider war. The Munich crisis was reported in excruciating detail by network radio. NBC alone broadcast 443 separate programs between September 10 and September 29, an average of three hours per day. As James Rorty remarked in *The Nation,* "American citizens with nothing else to do could have listened during most of their waking hours to the 'sound of history' as it poured out of their loud-speakers." He also noted that this marked a turning point in politics and diplomacy because "for the first time history has been made in the hearing of its pawns."

In England, the pawns could hear those happenings on portable radios, an immense variety of which were being sold in 1938. With war seemingly imminent, portables were a sensible purchase in England, for they could be used to receive crucial news in bomb shelters and in homes with failed power. In England the portable radio was fast becoming a necessity.

American radio makers, who took part in the World Radio Convention in April 1938, in Sydney, Australia, doubtless noted with interest the surge in portable sales in England. The strong feelings against portables held by major U.S. companies were beginning to soften, as potential profits loomed ahead. Portables could have been manufactured with off-the-shelf technology in 1938, and in fact a few were (Chapter 8), but the majors apparently reasoned that some new technology for portables— tubes in particular—would be an asset in the showroom. Salespeople could call attention to the "engineering breakthroughs" that had at last made possible a "practical" portable.

By the early 1930s, the development of new special-purpose tubes— even entirely new tube families—had become rather routine. In 1936, for example, engineers building home radios could choose from considerably more than one hundred different tubes. Although upgrading of battery tubes (like those used in farm radios and portables) did not receive a high priority, several new families of battery tubes had been created during the decade.

In August of 1938, Sylvania announced in *Electronics* "an important contribution to the radio industry," a new family of battery tubes (Fig. 9.1). The latter were especially well suited for portable receivers because of their 1.4 volt filaments (easily powered by dry batteries) and very low drain (all but one was .07 watt). However, by no stretch of the imagination could these battery tubes be called a breakthrough innovation. They represented the kind of mundane progress to be expected of good gray engineering. The battery tubes of 1938, then, were an incremental improvement on farm radio tubes in response to the expectation that Americans might be poised on the threshold of a new portable era.

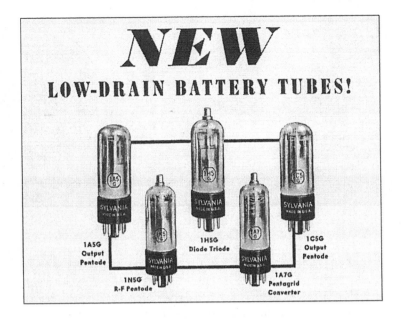

9.1 Sylvania's new battery tubes of 1938

On Halloween Eve of 1938, a few months after Sylvania began publicizing the new battery tubes, Orson Welles broadcast his unique adaptation of H. G. Wells' story, *War of the Worlds,* on the CBS Mercury Theatre. This thriller described a Martian invasion, treated like a newscast, which Welles placed in New Jersey. Although he prefaced the program with an announcement that the story to follow was a dramatization, many people apparently tuned in late and missed the warning. The result was near panic; two scientists from Princeton University set forth to observe the alien foe; in Newark, St. Michael's Hospital treated more than a dozen people for shock; emergency telephone lines were tied up throughout the northeast as people sought information, ambulances, and gas masks. (The Welles broadcast had longer-term effects by giving science fiction a boost. Not only were new magazines founded shortly afterward, but the first World Science Fiction Convention was held the following year in New York.)

The immediate reaction to Welles's broadcast above all testified to the hold of radio news on the American people. Because of radio reports of events in Europe, China, and Africa, the world had become a smaller—and more frightening—place. From more frequent news broadcasts arrived more vivid visions of America's future: another European war. The nation's young men would again be called to bear arms against Germany and her allies. With war imminent, could Americans afford to be away from their radios for very long?

According to radio makers, the answer was no. With Sylvania's new battery tubes, which became available in late 1938, companies large and small began to produce portables. Pushed hard in national advertising, these radios would make it possible for Americans, wherever they were, to hear world leaders and distinguished commentators chart the progress of the new apocalypse.

Many companies took credit in their ads for introducing this "entirely new kind of radio"—cryptohistory at its finest. Philco actually was the

PHILCO
A Musical Instrument of Quality

_____ THE TRAVELER

An entirely new kind of radio, invented by Philco engineers . . . the Philco 71T. Portable . . . self-powered . . . needs no aerial, ground or "house-current". Take it with you wherever you go . . . traveling, on trains, in hotels. Use it in camps, cottages, boats, at bathing beaches—anywhere indoors or outdoors. It plays, without "hooking up" to anything!

"She shall have music wherever she goes!"

9.2   Ad for the Philco 71T, first set of the portable radio revival, late 1938

first to sell its new portable, in the fall of 1938, but such a set was hardly new. What is more, as a small suitcase covered with airplane cloth, the Philco 71T (Fig. 9.2) was style-wise a dead ringer for a few of the earlier pseudo-portables. Nonetheless, Philco had close ties to Sylvania, and in all likelihood had asked Sylvania to design the tubes. Clearly, Philco did figure importantly in the portable radio revival, but the genre was not "invented by Philco engineers," as their ads claimed. (Predictably enough, Zenith rediscovered its 1924 Companion, which in company newsletters and annual reports became "the first portable radio.")

In the early months of 1939, more than a dozen other manufacturers joined Philco in the marketplace with their own portables; by late May at least twenty-eight set makers were busily cranking them out. By the end of the year, consumers could choose from more than 150 models (Fig. 9.3). Portable sales exceeded 200,000 by mid-1939, leading *Newsweek* magazine to proclaim the arrival of a "portable fad." By the end of that year, when Europe again went to war, 850,000 portables had been purchased along with more than eight million other radios—the most sold in any year up to that time. This time, obviously, the portable radio was a spectacular success.

Nearly every house in America had a radio by the time Germany invaded Poland in the autumn of 1939. When Great Britain and France retaliated by declaring war on Germany, Americans tuned in and held their breath. In the opinion of some historians, Roosevelt had already broken the Neutrality Acts by diverting war materiel to Great Britain, and he now called for these acts to be modified. Most people felt that it was only a matter of time until the United States would be drawn into the conflict. Americans continued to buy radios of every kind; no longer merely news, war had become, like soap operas, riveting entertainment.

Newsweek, in September of 1939, described the radio response to the outbreak of war: "Broadcasting studios moved suddenly into 24-hour emergency schedules, armies of correspondents and commentators were mobilized, and such a stream of war-and-peace confusion poured into living rooms that Americans were left almost as tense or groggy as a people awaiting an air raid."

To curb some of the excesses of war coverage, the three major networks working with a committee of the Federal Communications Com-

a

b

c

mission drew up a code, which was immediately adopted by the National Association of Broadcasters. Among its provisions was the following: "Every effort consistent with the news itself is made to avoid horror, suspense, and undue excitement." Though sanitized somewhat, war broadcasts continued to captivate Americans. And more portables were sold.

Though manufacturers may have been spurred to produce portables in the belief that Americans would never want to be far from a functioning radio, few ads stressed that the sets were ideal for receiving bad news from abroad. The emphasis, instead, was on the leisure activities that the portable could accompany. As a Philco ad in April 1939 suggested, "Take it with you wherever you go . . . traveling, on trains, in hotels. Use it in camps, cottages, boats, or bathing beaches—anywhere indoors or outdoors." By hammering on this theme, which almost all 1939 ads did, the portable radio was indelibly identified with happy times outdoors—a traveling companion for outings and vacations. By 1939, most middle- and working-class Americans did finally have paid vacations, and so could contemplate such a purchase.

The portable radio now provided a fine companion to leisure activities, and this use eventually came to be among its most important. Younger Americans, especially, took to the portable, as the following account in *Consumers' Research Bulletin* testifies:

> To the younger generation at least, the portable radio appears to have climbed almost into the necessity class . . . at the beach, at picnics, in the country, and even spectators' stands at sporting events, portable radios can be seen and (unfortunately, at times) heard. Sales of sets for use by men in military and naval training have also increased the number in use. So common have the small sets become that complaints against their use are beginning to appear in the newspapers.

On these new radios youthful Americans could listen to the music of the day from the "big bands." The bands of Benny Goodman, Glenn Miller, and hundreds of less famous personalities created the popular "swing" music, which was another make-over of jazz.

The intimate relationship between playtime and the new portables was emphasized visually by the sets' suitcase style (Fig. 9.3). The cases were usually made of fabric-covered plywood and had leather or plastic handles; they resembled small overnight cases, and some were actually manufactured by luggage companies. Although a few pseudo-portables and true portables had looked like luggage, such sets were quite uncommon. Thus, designers could borrow from them and create a distinctive style that almost everyone would perceive as new and unique to the portable radio. In this manner, portables were set apart in a visually dramatic manner from midgets with handles.

Compared to those of the twenties, this new generation of portables was inexpensive. Philco, RCA, Emerson, Zenith, and many other companies had sets for around $20—hardly more than a decent table model. Other portables, of course, could be bought for as much as $50. Although $20 was a week's wages for many working-class persons, it was the sort of discretionary purchase that one could save up for in a matter

d

e

9.3 A sampling of 1939 portables. *a,* RCA 94BP1 (12-3/8″ wide); *b,* Packard Bell 41 (11-7/8″ wide); *c,* Pilot T1451 (13″ wide); *d,* Sentinel 151BL (12″ high), closed and open views; *e,* Emerson CT275 (8-7/8″ wide)

of months. The portable radio was still a luxury item but affordable to many.

During the Christmas season of 1939, retailers expressed surprise at how well portables—the "summer" radios—were selling. Evidently, many were being bought as gifts for family and friends. The new portables conveyed the appropriate messages for a gift at that time, when America was emerging from depression. Not only did the portable radio connote leisure activities and discretionary income, but it was soon regarded by trendsetting consumers as a social necessity. *The American Home* magazine, for example, in July 1939, ran an article on the products needed "for a perfect summer." Among the twenty-seven upscale items, which included beverage glasses and tray, lawn furniture, and a "cherry corner cupboard," were *two* portable radios. The well-to-do family clearly needed a portable radio for outdoor entertaining. For families still barely making it, the gift of a portable radio, symbol of upscale leisure, advertised that better times were imminent.

Humorists had a field day with the new portables. A common theme was the playing of portables at inopportune times or in inappropriate places (Fig. 9.4). One memorable *New Yorker* cartoon showed a wedding party emerging from the church. The bride intoned the groom (who was carrying a portable radio), "Can't you put that damn thing down for even five minutes?"

Like small suitcases, the 1939 portables were a bit bulky, especially for pockets. Many measured around 12″ wide, 8″ high, and 7″ deep, commonly weighing fifteen to twenty pounds with batteries. According to Hugo Gernsback, writing in *Radio-Craft* in May 1939, the new portables had "much too great a weight for an ideal portable." Always impatient with the state-of-the-art, he urged radio makers to build sets weighing five or six pounds, because "I believe that there is a tremendous market for such a lightweight portable receiver." Actually, a few companies shared this belief, perhaps influenced as well by the parade of smaller portables built by the radio hobbyists.

In mid-1939 several more diminutive sets debuted, the most distinctive of which was made by Majestic ("Mighty Monarch of the Air"). Advertised as "the first *truly* portable radio you've seen . . . in a camera-

9.4 Portable radios were a subject for the cartoonist's art

"They're sparring in the center of the ring"   NED HILTON   "That's Professor Watkins—Current History."

9.5 Ad for the Majestic 130 camera radio, 1939

9.6 RCA's miniature tubes of 1940

9.7 Eveready 67-1/2–volt B battery for personal portables (3-5/8″ high)

style case," the three-tube Majestic 130 weighed less than four pounds (Fig. 9.5). Although tiny compared to run-of-the-mill 1939 portables, the Majestic 130 could not be placed in a pocket—not even a coat pocket. Yet, Majestic had won the shrinking derby. With the components and batteries available in mid-1939, further size reductions were simply not possible. To breach the limits and make a true "personal portable" (as Gernsback and the experimenters advocated) would require miniaturization of tubes, speakers, and so forth.

Recognizing in early 1939 that the portable revival had considerable momentum, RCA began work on another generation of battery tubes. These tubes were targeted for "personal" portable receivers, and so, for the first time, engineers set out to shrink the tubes employed in home battery radios. Like Sylvania's modified farm radio tubes of 1938, RCA's "miniatures" involved no technological breakthroughs. RCA had already been making tiny tubes since 1934 for short-wave applications (the "acorn" tubes), and so was able to draw on this expertise. Writing in their technical journal, RCA engineers stressed that the design of the miniature tubes had been simplified so that they could be manufactured by standard production techniques.

In November of 1939, RCA announced in *Electronics* its new family of four miniature tubes, which together could be used to make a superheterodyne receiver (Fig. 9.6). In view of the scant effort put into improving—much less miniaturizing—the portable radio in the 1930s, it is ironic that the personal portable was the immediate impetus for the new tubes. Perhaps this is not so surprising in view of who was still president of RCA: David Sarnoff. Miniature tubes helped to realize his early vision of the personal portable.

Throughout the 1940s and 1950s, hundreds of new miniature tubes were created for countless applications. Tube radios of the 1950s and 1960s used them almost exclusively; and they were quite common in post-war televisions. This was an important technology that originated in portable radios.

Like the last farm radio tubes, the miniatures had a 1.4-volt filament, and so could be powered by dry cells—even a flashlight D cell; they were also low-drain, .07 watt. However, their plate requirements were less; they could operate on 45 volts, though 67.5 or 90 volts produced greater volume.

As of the late 1930s, B batteries (for plate voltage) had not been miniaturized. However, making smaller batteries was not hampered by any technological barriers. Batteries could be made ever smaller, but there was a price: smaller batteries, having less current, ended up costing more per watt-hour of usage than large ones. Smaller B batteries were rapidly made for the personal portables by Eveready (Fig. 9.7).

Additional miniaturized components were also quickly produced by suppliers as radio firms made plans to manufacture personal portables. Tuning capacitors, speakers, and intermediate frequency transformers were reduced as never before. Ironically, in late 1939 and early 1940, the personal portable was on the cutting edge of radio miniaturization. In all of electronics at that time, only the hearing aid drove miniaturization harder (see Chapter 12).

In April of 1940 Sonora Radio and Television Corporation introduced a "midget portable," its "Candid" model (Fig. 9.8, above). Sonora was followed closely by RCA and a half dozen other manufacturers who also made personal portables using the new miniature tubes. Tens of thousands of these compact sets were sold very quickly in New York City, a test market for RCA, where household radio ownership already exceeded 98 percent. The surge of portable sales in that metropolis clearly highlighted the premium that people placed on being able to hear radio anywhere. It is easy to appreciate that someone with such a radio on the street, in a bus, or at the office would attract a certain amount of attention as the dispenser of an important resource—access to world events. The price of this instant popularity was $20.00, the cost of RCA's first personal portable, the BP10 (Fig. 9.8, below). Some dealers eased the burden of purchase even further by offering Christmas lay-away and dollar-a-week installment plans.

Not surprisingly, the major style adopted for the personal portable was that of the camera (Fig. 9.9). Camera radios had cases of metal or the "material of the future," plastic. On the metal sets (like the RCA BP10 and the Motorola A1), a plastic lid or back was needed to accommodate an antenna. Incidentally, plastic and metal cases were able to express the streamlining style much better than the larger luggage sets. The futuristic camera radios were rather popular, and many were given to young men entering the armed services.

Though we may find it hard to believe today, plastic was the miracle material of the 1930s. It was used in an increasing variety of household items, and the radio industry was in 1940 its biggest customer. A handful of camera-style sets used some of the newer plastics, which had wonderful names like Lumarith and Lustron. Though lighter and less brittle than bakelite, these plastics had a tendency to shrink, especially when exposed to heat (from a rectifier tube or even sunlight). Many of the magnificent streamlined cases became warped and cracked, and the reputation of the new materials was soon sullied.

Personal portables really were quite compact and light—for their time, weighing (with batteries) only five or six pounds. Most of the camera-style radios could fit into a large overcoat pocket, but only a few models were actually promoted as "pocket-size." The Emerson 432 (from 1941) was a triumph of clever design, weighing but three pounds and taking up less than eighty cubic inches (8-1/2″ × 4-1/2″ × 2-1/8″ maximum dimensions). Advertised as the "world's smallest and lightest," this radio was easily swallowed up by a big purse or a small coat pocket. The Emerson 432 opened up (and turned on) like a music box (the same as the RCA BP10), and so could not be played while in the pocket.

The camera design, of course, had the same meaning as the small suitcase. Both loudly proclaimed that the portable radio was the perfect accompaniment for travel and outdoor leisure activities. Camera-style portables acquired another meaning as well, but not as positive: they were battery hogs. Typically, a set using one D cell for the filaments (the A battery) would have to be opened up for battery replacement after every three to five hours of use. Portables employing two D cells could be left alone for ten to fifteen hours. At five to ten cents each, the A batteries of

9.8   The first personal portables to use miniature tubes, 1940. *Above,* Sonora Candid; *below,* RCA BP10 (8-7/8″ wide)

a

b

c

d

e

9.9 Other camera-style personal portables, 1940–41. *a,* Philco 89C, 1940; *b,* Zenith Companion Poketradio 4K600, 1941; *c,* Admiral Bantam 29-G5, 1941 (8″ high); *d,* Motorola A1, 1941; *e,* GM 975775, 1941 (7″ high)

9.10 RCA battery pack for larger portables (9-5/8″ wide)

personal portables put a big drain on the pocketbook. B batteries—which cost about $2.50 for the small sets—lasted much longer, typically forty to sixty hours. The average cost for running a camera-style portable was about six cents an hour, at a time when five cents could still buy a cup of coffee. Many people surely regretted their purchase of a personal portable when the battery costs started mounting.

The larger portables had much lower operating costs, about one or two cents an hour for the most economical models. A set of A and B batteries, sometimes available in a single battery pack (Fig. 9.10), ran about $3.75. This was a hefty outlay, but it was only needed every two hundred to three hundred hours; one battery pack could easily last for an entire summer. Even so, radio makers sought new ways to improve battery economy. For example, a few (like the Automatic P1) had a provision for recharging the batteries, which could ideally extend their life by a factor of two to four.

In 1941 General Electric solved the battery problem more definitively by building a portable of radically new design that used a two-volt storage battery, which would last several years. The General Electric LB530,

with built-in battery charger, was similar electronically to a car radio.
The set cost $39.95 and, with its complex aluminum housing, weighed
only 16.5 pounds. A new Willard battery was $6.50.

The single greatest hazard to ordinary batteries was a radio acciden-
tally left on. To reduce this risk, set makers sometimes employed clever
mechanical devices that indicated visually when the set was operating.
For example, in the Motorola 41D, the tuning dial came into view only
when the power was turned on (Fig. 9.11). It is doubtful that these aids
saved many batteries, but they would have been impressive in the
showroom.

One feature that made portables so attractive, eventually, was that
many could be used like a table model, that is, plugged into house current. These were called three-way (AC, DC, battery) or convertible sets.
Only a handful of three-way sets were made in 1939, but they became
rather common as the portable revival progressed. In 1941, even some
camera-style sets had three-way capability (Fig. 9.9c).

Three-way capability was achieved by the addition of a rectifier tube
that was deployed, usually automatically, during AC operation. This tube
and accompanying filter capacitors supplied direct current for plates and
filaments of the remaining tubes.

The convertibles opened up new possibilities for use of the portable.
In particular, a family contemplating the purchase of a table model might
be inclined instead to buy a three-way portable. For a few dollars more
than a cheap midget they got a radio that could also be taken on a picnic
or to the beach. Thus, another reason for the portable's success was that
people were buying more radios of all kinds, including table models, and
three-way portables were in effect a very flexible table model.

The few farm families that still lacked radio might have turned to portables because they were cheaper than most farm radios. A *Life* magazine
ad of early 1939 showed an RCA "Pick-Me-Up" portable being used by a
farm family, highlighting what probably became an important market for
the new battery sets.

Once the portable radio had established itself as a viable genre, designers gradually were given greater freedom to experiment with the
portable's outward appearance (Fig. 9.12). By late 1941, airplane-luggage
cloth had become passé, though it was still used widely. A variety of
other fabrics was also used, including imitation leather. Many larger sets
departed from the luggage style, adopting a trimmer, lighter look that
seemed to suit the portable (Fig. 9.12e, f, g). On the other hand, a few
large wooden sets became bulky and bizarre. A case in point is the Philco
844T, which resembled a miniature roll-top desk (Fig. 9.12c). Clearly, a
distinctive new style for the portable radio had not yet been achieved.
Designers were obviously groping for other conventions but their efforts
were cut short by the war.

On December 7, 1941, President Roosevelt told the nation over radio
that Japan had just bombed Pearl Harbor. Within twenty-four hours,
Congress declared war on the Axis powers.

Before the United States government ordered a halt to civilian radio
production in April 1942, almost four million portable radios had been

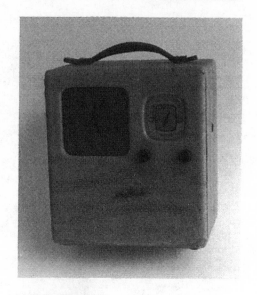

9.11 Motorola 41D, of 1939, with "Battery
Off" indicator

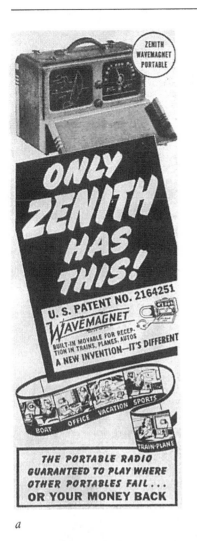

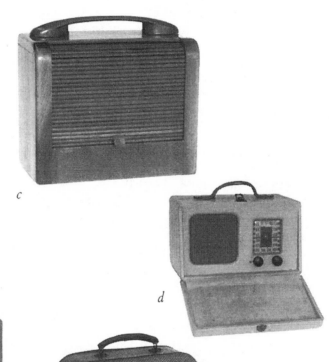

sold. During this brief period of popularity (just three-and-a-half years), tube portables achieved a high degree of electronic perfection.

Although portables were a great commercial success, not all Americans took part in the portable "craze." Indeed, at the outbreak of the war, most American families did not own a portable radio despite their stylish exterior, upbeat meaning, and respectable performance. Some people had to have news everywhere, and others required musical accompaniment on picnics. As flexible table models and farm sets, still more portables found a place in the American home. Clearly, portable radios had become much more than a plaything for the rich, but the majority of families still did not regard this wonderful product as a necessity.

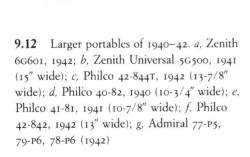

9.12 Larger portables of 1940–42. *a,* Zenith 6G601, 1942; *b,* Zenith Universal 5G500, 1941 (15″ wide); *c,* Philco 42-844T, 1942 (13-7/8″ wide); *d,* Philco 40-82, 1940 (10-3/4″ wide); *e,* Philco 41-81, 1941 (10-7/8″ wide); *f,* Philco 42-842, 1942 (13″ wide); *g,* Admiral 77-P5, 79-P6, 78-P6 (1942)

One of the last portable radios to enter production before radio factories were harnessed for Uncle Sam was Zenith's Trans-Oceanic. This was an extraordinary radio that had five shortwave bands (in addition to the broadcast band). Although only a few tens of thousands were built, in far-off lands the Trans-Oceanic brought joy and comfort to American soldiers.

The story of the Trans-Oceanic begins with none other than Eugene F. McDonald, Jr., trendsetting consumer and Zenith's president. It was his custom to take short vacations on Lake Michigan, cruising in his yacht. On one of these excursions in the summer of 1939, McDonald had along a standard Zenith portable. Being some distance from civilization, however, he was unable to receive a single station. On August 2 he sent a radiogram to Zenith's chief engineer back in Chicago, instructing him to begin work on a high-performance portable. Howard O. Lorenzen was an engineer assigned to the new project, and he recalls that McDonald "was one who wanted to be abreast of the latest news and world developments, and so this lack of daily news was very irksome to him." The solution was found in shortwaves, an area that Zenith had pioneered in the early twenties. A sensitive radio with shortwave bands could pick up European news broadcasts directly.

During the next two years, Zenith's laboratory supplied a succession of Trans-Oceanic prototypes—about twenty in all, which McDonald himself tested in the field. He was a severe critic who paid attention to every detail; the radio had to work well before he would authorize production. When he was finally satisfied, McDonald furnished his friend Admiral Donald MacMillan two of the latest laboratory receivers for further trials. After encountering a few glitches, MacMillan was able to give "Gene" a glowing report from Greenland on July 27, 1941: the radio worked "beautifully"; they got "both European and American stations day and night and are keeping fully informed as to what is going on both at home and abroad."

Shortly thereafter, McDonald put the Trans-Oceanic Clipper (Fig. 9.13) into production, against the advice of other Zenith executives who felt that the set would be too expensive to find much of a market. McDonald was equally sure that people would buy the radio because it performed so well. Being the boss, McDonald prevailed; he was also right. When the Trans-Oceanic assembly line shut down on April 22, 1942, it was the highest priced U.S. portable (over $100), but many thousands of orders remained unfilled.

The United States was already at war during the three-and-a-half months that the Trans-Oceanic was available to Zenith dealers. Although portable radios were useful for Americans stateside seeking war news, the Trans-Oceanic was especially useful to the soldiers abroad seeking entertainment and news about home. During the war Zenith received scores of touching testimonials from American servicemen. Typical was the comment of Private James Henry Brown, writing in 1944: "I now own one of your greatest creations, the Zenith trans-ocean portable deluxe model 7G605, and it has been of great service to me and a lot of the boys up here in the Aleutians. It is constant entertainment. With six bands to tune, we can get programs all times of the day."

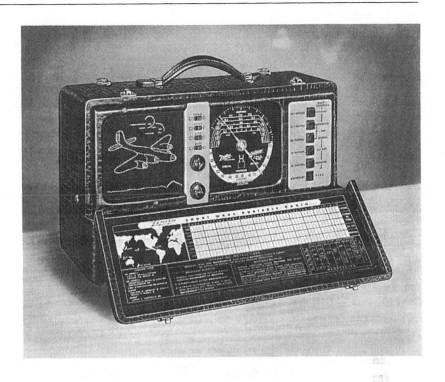

9.13 Zenith Trans-Oceanic Clipper 7G605, 1942

Other letters described Trans-Oceanics that kept on playing even after splashing into the Mediterranean, being bombed on a Pacific island, or falling off a fast-moving truck in Africa. Such letters were published in the pages of the *Zenith Log*, the employee newsletter, under headings such as, "Zenith Folks, Read This—if you want to be proud of your work." To Zenith employees, the Trans-Oceanic doubtless was a source of great pride, which reinforced loyalty to the company.

Although the Trans-Oceanic was out of production for the remainder of the war, McDonald had set aside a secret stash of the sets for his friends. Periodically, he bestowed one of these "masterpieces" (McDonald's term) as a gift. Frank Sinatra, for example, received one at a public presentation before more than 1,000 admiring phonograph record clerks. The use of the Trans-Oceanic portable as a conspicuous gift to a celebrity enhanced its desirability, especially since mere mortals could not buy one at any price.

Still other uses were envisioned for the Trans-Oceanic portable back in the states. The Trans-Oceanic was featured in a *Radio News* article of May 1942, entitled "Radio for blackouts and power line failures." This piece described the need for a "family refuge" room in the event of air raids and provided advice on stocking it. After food, of course, the Zenith Trans-Oceanic was the refuge room's most important furnishing.

The war brought virtually full employment to America, but civilian sacrifices were many. Military needs created shortages that led to rationing of gasoline and other necessities. Tires, stockings, refrigerators, washing machines, and even shoes all were in short supply. Batteries for portable radios were sometimes quite scarce (Fig. 9.14, below). In one ad for Eveready radio batteries in November 1942, readers were alerted in a small box to the problem: "Naturally, the Services came first. At the time this magazine went to press, however, there was available for civilian use a limited supply of 'Eveready' 'Mini-Max' batteries."

In that same year radio servicemen were being provided with tips on how to convert radios to use the few types of battery that would be available. (Larger and odd-sized battery packs were not being made at all.) Later in the war, advice would be given on how to convert a battery-only set to AC operation.

Radio tubes were also scarce, as tube factories changed over to military production. Doubtless many a radio was silenced for lack of a replacement tube. But servicemen were creative, scavenging tubes and other parts from older sets ("junkers"). Sometimes a set could be rewired slightly to accommodate a tube that was similar to the defective one.

Even during the war electronic tinkerers continued to design small portables. For example, in *Radio for the millions*, a do-it-yourself project book published in 1943 by *Popular Science*, plans were provided for a "pocket receiver for sports fans." The set had one acorn tube and employed Armstrong's super-regenerative circuit; it was limited to headphone operation. The loop antenna was wound on the exterior of the case, which had been fashioned of cigar-box wood and was very tiny ("little larger than a tobacco tin"), easily sliding into a jacket pocket. Tuning and volume controls, located on the top of the radio, could be adjusted while the set was in the pocket. Immediately after the war, a pocket receiver reminiscent of the *Popular Science* sportset would be produced commercially (Chapter 12).

As in the First World War, portable radios had real work to do for the military. The availability of miniature tubes and other diminutive components made possible military transceivers of unprecedented portability. Two kinds of portable radios were carried by troops on the front lines: Handie-Talkies, which were small (3″ × 3″ × 12″) and could be easily held in the hand, and Walkie Talkies—somewhat larger (17″ × 12″ × 7″)—carried on the back.

World War II was the last American war in which civilians also suffered significant privations. All the major consumer-electronics manufacturers were churning out supplies and equipment for the military. Tubes, for example, were being made in 1944 at a rate of 400,000 per day. The trendsetters had to sit on their hands; there were no new goodies to grab. Radio companies were aware that pent-up demand for sets was growing. Sylvania predicted that within five or six years after the war, a total of 100 million new radios and televisions would be sold. (The true number was pretty close.) To help assuage the consumer's frustration at having money but nothing to spend it on, radio company ads promised better things ahead. The new technologies developed during the war, consumers were assured, would have dramatic spin-offs for the ordinary consumer.

In *Radio-Craft* of April 1945, the president of Galvin Mfg. Co. (Motorola) claimed that "when peacetime production of radios is resumed, our new line of AM and FM radios will include many . . . war-time improvements." As Sarnoff himself put it in 1944, "When peace comes it

"They're rationing sugar in the States. What the hell is sugar?"

"Even with all his money, he couldn't buy a new battery, so he had to get an electric eel."

9.14   Wartime cartoons poke fun at rationing, shortages, and the portable radio will find, as it has at the end of every war, new inventions awaiting to be applied to everyday life." These claims were not unreasonable; World War I, after all, had led to advances in radio apparatus and in the mass-production of components.

Some of these promised spin-offs did take place, but only one war-born device—radar—was to have a truly profound influence on post-war consumer electronics. Radar, predicted by Gernsback in 1911, was perfected by both sides before and during the war. Radically new tube types were needed to generate the ultra-high frequencies used in radar. Such tubes would become important later in television and FM broadcasting—and in the microwave oven.

Detection was also a serious problem at radar frequencies, and vacuum tube detectors proved to be inefficient. Engineers recalled the venerable crystals used in early radio, and so the government mounted a massive research program to produce crystal detectors for radar. The challenge was to make crystals having uniform, predictable properties. Government-supported research was undertaken in many university labs on the two most promising materials: germanium and silicon. Their electrical properties were thoroughly studied and techniques were refined for producing crystals of astonishing purity. Not only did these efforts lead to reliable detectors for radar, such as the "germanium diode," but they established a foundation for the invention of the transistor (Chapter 12).

Cathode ray tubes (used in radar displays) were similar to the video tubes of television. Many companies developed techniques for mass-producing the radar tubes; when the war ended, these factories were readily converted to making "picture" tubes for television. Other civilian applications of radar, predicted by Hugo Gernsback in 1945, came true only partly: he foresaw that radar would be used to prevent collisions in airplanes *and* automobiles.

Writing in early 1944, in *Radio-Craft*, Gernsback also had a prediction about miniature radios: "A complete, self-powered radio set that will be no larger than a dime is not an impossibility and sooner or later it will be produced." A few months later, he devoted an entire editorial to "Miniature Radios." "The Pocket and Vest Pocket type of radios," he said, "fill a real demand." Gernsback was on a roll: "I predict that before long many millions of these radios will be built annually. An entirely new art will be reared upon them. They will be built mostly by female workers who are more nimble in assembling the exceedingly small parts than men." Such radios, he continued, would be "immediately acceptable to the public at large. It becomes an article of universal demand by everybody. It appeals to all ages, including children. A radio set that fits the hand, that is light and compact and which can be bought for about $10 becomes irresistible to the masses." Gernsback also anticipated that such radios would be made in a variety of forms, such as books, dolls and animals, perfume bottles, and in combination with clocks and cameras.

Needless to say, Gernsback was not the only techno-mancer of World War II. In a 1944 issue of *Radio-Craft,* various authorities predicted marvels such as push-button telephones, radio-controlled lawnmowers, and electronic speedometers for cars. Some enlightened prophecies never came to pass—for example, that AM radio would become obsolete because of FM's technical superiority.

Radio in the war years assumed an even greater importance in the lives of ordinary Americans. In addition to favorite programs was fresh news from the front. As David Sarnoff put it, "a home without radio is out of tune with the world." Radio was America's lifeline to far-off lands where friends and relatives, perhaps a member of the immediate family, were involved in the business of war. From fly-speck islands in the Pacific or obscure villages in France, foreign correspondents told their tales to a country that eagerly awaited every word. Vietnam was not the first American war to enter the living room through an electronic medium; in the sixties, though, it was hard to convince a seeing (and hearing) public that very much of the horror—visible on color TV—was really heroic. Radio correspondents during World War II presented war in a very different light.

In Europe, the war finally ended on May 7, 1945, when the Germans surrendered unconditionally. A few weeks before, President Roosevelt, re-elected the previous fall to an unprecedented fourth term in the White House, died suddenly. The silver tongue of radio had been forever stilled. Harry S. Truman became the nation's thirty-third president.

Though Germany and Italy had fallen, Japan battled on through the early summer. The final blow came in August, when U.S. pilots dropped atomic bombs on two Japanese cities. The atom bomb represented a technology gap of immense size that the Japanese could not soon close; the war was over.

When the mushroom clouds cleared over Hiroshima and Nagasaki, a chastened world could contemplate peaceful uses of atomic power. In October 1945 Gernsback joined a multitude of techno-mancers in proclaiming that "*the unlocking of the atom will change everything on earth. Future man will have for the first time an abundance of cheap power which can be used instantly wherever it is wanted.*" More extreme postwar proponents of nuclear power even claimed that electricity would become too cheap to meter.

In the meantime, Americans prepared for peace and, especially, for prosperity. Factories converted to civilian goods that were already on the drawing boards. As the consumer society rebounded with vigor, portable radios would find new places in American life.

10

A Parade of Post-War Portables

GIS RETURNING HOME FROM THE FOUR CORNERS OF the globe found a slumbering giant. The consumer society, sluggish in the thirties and dormant during the war, was ready for a spending orgy. Though taxes took a big bite from the wealthy, the working and middle classes (as well as farmers) had more discretionary income than ever before. In 1946, the average factory worker made $2,230, spending only $1,278 on food, housing and utilities, furniture, and clothing. Americans had been promised a future of new homes and cars and countless conveniences, and that future was now. Post-war Americans were thinking of things—new things, big things. In "Dreams of 1946," *Life* magazine showed a front yard filled with fresh products—from televisions to dishwashers—coveted by the average family.

Manufacture of most consumer goods resumed within a year after the war's end, though supply often lagged considerably behind demand. The resultant inflation was a problem for a few years, but the real purchasing power of the ordinary American continued to grow. A cornucopia of new products tempted consumers in lavish department store displays, and radio and television ads exhorted would-be buyers as never before. More than thirty magazines reached a million or more homes each, and in their pages colorful ads hawked everything from Prince Albert pipe tobacco to the Stinson "personal" airplane—not to mention the Westinghouse "electric sink." Following a new home and car, the next most momentous purchase was probably a television.

To meet the anticipated demand, the major radio companies lunged headlong into the television age. In addition, an enormous entrepreneurial fringe of TV makers sprang up quickly: more than 125 companies were cranking out table models, consoles, and swollen TV-radio-phonograph combinations. Television manufacture obviously made it possible for many electronics firms to convert briskly from military to consumer goods.

Sarnoff's sense proved right again: television was rapidly embraced by a novelty-seeking consumer society that was already hooked on electronic entertainment. Despite *Life* magazine's December 1947 dig that the programming fare of the new medium was "mediocre to bad," almost a million sets were sold the following year. Programs did improve—somewhat—as network radio was plundered for the best talent. In 1950, television had already entered ten million homes—at an average price of well over $200. Just four years later, the number of families with TVs exceeded thirty-five million. Penetration of televisions was so rapid that sales leveled off by the mid-fifties; in the last few years of the decade—during a recession—sales actually slumped.

Most of the TVs bought in the late forties and early fifties were huge consoles that squatted, sometimes majestically, in a prominent place in the living room. Even in the late forties, however, quite small sets were available, including a seven-inch, thirty-one-pound Sentinel advertised—for $206—as a portable. It did have a handle, but could not run on batteries.

As console sales peaked in the mid-fifties, most manufacturers began to offer "portable" televisions. Though relatively compact and inexpensive (some were under $100), most weighed at least forty pounds (an interesting exception was a Hotpoint set weighing fourteen pounds) and all ran on house current only. But, like the early Sentinel model, they did have handles. The similarity of these portable televisions to the pseudo-portable radios of the Depression is striking. In both cases, the "portable" label distinguished the lighter, more easily moved models from their heavier, immobile relatives. And both influenced the style of true portables that appeared later.

Television, of course, was a pre-war technology, which finally came to fruition in the marketplace. The years between the end of the Second World War and the coming of the first transistor radio (late 1954) saw vast leaps in electronic technology that led, eventually, to entirely new home entertainment products. (The story of the transistor and its application to portable radios is postponed to Chapter 12.) It is in the context of two such new electronic wonders—color television and FM radio—that the portable radio continued to evolve and play a part in American life. Like black-and-white television, FM and color television had pre-war roots, and telling their stories requires some backtracking.

Color television was a technology that had long entranced the techno-mancers. It was, after all, a logical extension of television, in the same way that color photography was the next step after black-and-white. By the late thirties, when color movies came to American theaters, color television had become a cultural imperative with a constituency that included major companies.

The emergence in the late forties and early fifties of a practical system of color television illustrates the role that corporate rivalry can play in stimulating technological innovation. In this case, the competition (which dated back to the late twenties) was between communications giants RCA and CBS, and it began at the top of the corporate ladder, with Sarnoff and Paley. CBS engineer Peter C. Goldmark in his autobiography, *Maverick Inventor,* described the atmosphere at CBS in the late thirties: "The urge to beat RCA and its ruler, David Sarnoff, was such an overriding force at CBS that it actually began to shape the direction of my own career."

In March 1940, Goldmark attended a showing of *Gone With the Wind*. He was enthralled, not by the story or the performers, but by the vivid, realistic color: "It was the first color movie I had seen, and the color was magnificent." Throughout the long movie he became "obsessed with the thought of applying color to television." Soon afterward he created for CBS a system that would eventually be known as "field-sequential" color TV; detractors, however, called it "Goldmark's whirling dervish." This latter label was not entirely unjust because both the camera and receiver required a spinning color wheel. Like early black-and-white television, the CBS color system was partly mechanical, though (if we can take Goldmark's word for it) the picture was large and did not flicker—and the color was superb.

At the first industry meeting to discuss black-and-white TV standards in 1940, Goldmark discomfited the crowd by announcing CBS's color work. In September the new color system was successfully demonstrated to the Federal Communications Commission (FCC). RCA believed that its own enormous investment in black-and-white television was clearly at risk: if CBS succeeded in getting color television approved at this early date, then consumers might simply leapfrog black-and-white. So Sarnoff and RCA, for so long identified with electronic innovations, staunchly opposed color television. Before the color television war could begin in earnest, however, a real war intervened.

In 1947 the FCC turned down CBS's request for color television, parroting the RCA line that the system was "premature." Another factor that influenced the decision was the incompatibility of the CBS system with existing black-and-white receivers (the latter would have become obsolete if the CBS system had come into general use). This decision was quite predictable on other grounds as well: throughout the industry, companies had tooled up to produce black-and-white sets. They did not want the game changed before they had any winnings. Goldmark persisted nonetheless, improving the CBS system. In the meantime, Sarnoff covered all bets by assigning RCA engineers the task of bringing about an all-electronic system of color TV. If color was really in the cards, RCA would not have to pay to license CBS technology.

In September of 1950, RCA, CBS, and one other company demonstrated their color systems to the FCC. The headline in the next day's *Variety* said it all: "RCA lays colored egg." In contrast, Goldmark's system performed flawlessly, and so CBS was granted a commercial license to march ahead with color TV. With their newly acquired manufacturing arm, Hytron, CBS was preparing, with color TV, to become an electronics giant.

Publicly humiliated, Sarnoff returned to RCA and ordered his engineers to redouble their efforts—money was no object. The Korean War bought Sarnoff all the time he needed. RCA engineers developed an all-electronic color TV using a three-gun picture tube. In 1953 the FCC reappraised the status of color TV and reversed its earlier decision, approving an all-electronic system. Apparently Sarnoff's argument had been accepted, that the CBS system "would saddle an all-electronic art with a mechanical harness." By this time, of course, RCA and many other manufacturers, having recouped their investments in black-and-white, were ready to add color TV to their lines.

RCA's victory over CBS (that is, Sarnoff's victory over Paley) came at enormous expense. It is estimated that RCA spent at least $130 million to create color television, a sum that was not repaid by profits until well into the 1960s. In 1954, RCA and a few other companies offered to the public the first color sets—at a breathtaking $500 and up.

Zenith did not put color TV into commercial production until 1961. According to McDonald, still at Zenith's helm, RCA's three-gun picture tube was "a Rube Goldberg contraption if ever there was one," a judgment in which many others in the industry concurred. McDonald went further, accusing RCA of using "the American people as guinea pigs in order to . . . sign up its patent pool licensees for another term." In the meantime, Zenith invested millions trying to build a better picture tube, which the company finally did in the sixties, though it still had three electron guns.

Beginning in 1955, most other set makers brought out color TVs using the RCA picture tube. The hope was that color sets could compensate for flagging sales in black-and-white consoles and table radios. Color television was promoted lavishly, but the results at the cash register were disappointing. Only 25,000 sets were sold in 1954 and 1955; from 1956 to 1960, only twice did sales exceed 100,000 sets. Clearly the trendsetters were there, but few others were ready to buy. At the end of the decade, less than 1 percent of American homes had a color TV. By 1960, only RCA and a couple of other companies were still selling color TVs. Though a cultural imperative of some antiquity, color TV was a flop in the marketplace.

There were many reasons for color TV's failure to catch on, not the least of which was price. Throughout the fifties and into the early sixties, the cheapest color TV was still $500, the average price around $1,000. Another problem was that the complex sets broke down frequently and service costs mounted fast. Even when the receiver was working, the picture still might not be great; often the colors were perfectly registered only at the center. The folk wisdom of the time was that color TV "had not been perfected yet." Consumer magazines agreed. As late as 1960, *Consumer Reports* concluded "that color TV is only slightly more 'ready' now than it was in November 1956."

The biggest handicap, though, was more fundamental: there was nearly nothing to watch. Even NBC-TV, owned by RCA, had only a few hours of color programming a week in the fifties. Before each color program, the NBC peacock spread its polychrome wings to announce the momentous occasion. CBS had the odd color show, but ABC never broadcast in color. One industry executive summed up the dilemma: "like the sponsors who are unwilling to pay a premium for color telecasts without an audience, the audience is unwilling to pay a premium for a color television receiver without good color programming." American families had just bought a black-and-white television, which worked pretty well; there was little reason yet to upgrade.

In the mid-sixties, as their black-and-white sets aged, people began to notice that color TVs were getting better and that the number of color shows was skyrocketing. Consumers finally turned to color TV in large numbers; by the end of the sixties, more than twenty-five million American homes had them. At last, RCA made a bundle on color TV. But Zenith's go-slow strategy had paid off too—they were the industry leader. CBS was not even making televisions any longer. The technology had come a long way since 1940; by the late 1980s, of course, one could buy pocket color TVs—the ultimate extension of the portable radio.

In contrast to color TV, a product of clashing corporations and large industrial laboratories, FM was mainly the handiwork of one man: Edwin Armstrong. Although radios of the late twenties and early thirties were easy to use and had a rather pleasant sound, the last radio gremlin—static—still haunted reception. Among engineers and the radio-buying public, a sure-fire remedy for static had become, by the middle twenties, a pressing cultural imperative. Sarnoff himself had envisioned an attachment that would eliminate static from existing radios. Countless devices were advertised as static eliminators, though none really did the job. By the late twenties, most radio engineers despaired of ever solving the static problem.

This was precisely the sort of challenge that appealed to Armstrong. At the same time that superheterodyne radios were chalking up amazing profits for RCA (and for Armstrong, RCA's largest stockholder), the ingenious circuit designer and Columbia University professor was wrestling with a way to free radio from static and other interference. He soon came to appreciate that static was inherent in AM broadcasting. A solution more radical than a static eliminator would be needed.

From 1928 to 1933 he experimented with wide-band frequency modulation (FM), an entirely new system for impressing a signal (voice or music) on a carrier wave. Mostly on the basis of theory, engineers at AT&T and RCA had previously pronounced FM unworkable, but after five years of effort, Armstrong achieved success. Not only was the new system just about impervious to static (rejecting better than 99 percent), but it was nearly noiseless. Virtually all the sounds heard by the human ear— from the lowest rumble to the highest tweet—could be received on an FM radio without distortion. Thus, FM could become the basis of true high-fidelity radio.

Armstrong demonstrated the new system in 1933 to his old friend David Sarnoff. The president of RCA was properly impressed. However, with RCA's stellar earnings from AM broadcasting (through NBC) and huge investments in television research, he was unwilling to make a major commitment to FM. Armstrong's "static eliminator" was just too radical. (Later, Sarnoff's own beloved television would put the AM networks into eclipse—the story is in Chapter 13.)

Armstrong was confident that interest in FM would be high—at least among music-loving people, so he proceeded on his own. Because his earlier inventions had made him a multi-millionaire, Armstrong was able to pursue his vision of high fidelity radio without corporate (or government) support. In fact, the established AM interests (NBC and CBS) were hostile to FM, because it would mean new stations and new networks having vastly superior sound. The very existence of the AM empires was profoundly threatened by Armstrong's invention.

In his own laboratory with staff paid out of his own pocket, Edwin Armstrong perfected practical FM apparatus—transmitter, receiver, the works. On July 18, 1939, he began regular broadcasting on the world's first FM station, WZXMN, in Alpine, New Jersey. It had cost him more than $300,000. The first FM receivers for the home were made that same year by General Electric, and soon other companies, including Zenith, were manufacturing FM radios under license to Armstrong.

By late 1939, the FCC had received about one hundred and fifty applications for FM stations, and new channels were provided. Taking notice of this activity, in the fall of 1940 RCA offered to buy all of Armstrong's FM patents for a paltry $1 million. He refused to sell. By the time the United States entered World War II, more than forty FM stations had begun regular broadcasting and about half a million FM receivers had been sold.

Immediately after the war, FM was to suffer some major setbacks. The FM band authorized by the FCC in 1940 had been 42 to 50 MHz. In 1945 this band was reallocated to television by the FCC, and the FM band was moved to its present position, 88 to 108 MHz. A half million pre-war FM radios had become instantly obsolete. The FCC also cut back the power of FM stations. Armstrong was powerless to prevent this treachery, which had been justified on the basis of bogus technical grounds. FM survived in the post-war period, but it did not reach its potential, much less pose a threat to network AM radio.

Ironically, FM did thrive in television, where it supplied the audio. However, RCA and many other companies that employed FM in radios and televisions refused to pay Armstrong the royalties due for the use of his system. Armstrong filed suit and the resultant legal wrangles exhausted him, financially and mentally. He eventually won in court, but not before his suicide in 1954.

FM radio survived its creator, though it made only modest headway into American homes in the late forties and fifties. For most listeners of popular music, static-laden AM was adequate. About 10 percent of the table models sold during this period had an FM band; some expensive television-radio-record players also included FM. For classical music enthusiasts, of course, FM was the only choice.

The portable boom resumed after the war, pretty much where it had left off. All the major radio makers brought out a line of portables, and they were joined in the marketplace by an energetic entrepreneurial fringe. Indeed, more than fifty firms manufactured portable radios in the late 1940s, not all of them especially memorable (radios or companies). Names like Signal, Jewel, Dynavox, ECA, Chancellor, and Remler were among the radio makers of that time, some of which were defense contractors in the New York City area struggling to find civilian goods whose manufacture would be as lucrative as war materiel. Despite prosperity and strong interest in portables, few of the newcomers were still making these sets in 1950.

Like most consumer products, the portable radio was restyled after the war. Machine Age streamlining fell into disfavor, partly because of its association with the hard times of the Depression. That style had promised a new age, which was now; and so the new age needed new styles. Streamlining did not disappear altogether; it became a minor style among many other minor styles.

In the post-war period, designers seemingly were allowed free reign to forge new fashions that would suit the age of affluence. No product illustrates this anarchy better than the portable radio. Between 1946 and 1950, designers no longer had to hew to the pre-war formulas for portable design—small suitcases and cameras—because the portable radio itself had come to mean "traveling companion." These conventions were supplemented with new types that defy simple description. Portable radios acquired new meanings as designers played with style and people played with portables.

The years between the end of the Second World War and the coming of the transistor radio (in late 1954) represent a high-point in portable radio design using tube technology. These sets were not only reliable, but attractive, stylish, even whimsical. A creative freedom came to be expressed and embodied in the portable radio, reminiscent of 1930's table models. Never before could Americans choose from such a panoply of playful portables. Portable radios meant good times, and now it really showed—in wood, metal, and plastic.

Portables in fabric-covered wooden boxes were still being made in large numbers during the late forties, but for the most part their days were numbered (Fig. 10.1).

Leather was a popular material for post-war articles, especially women's clothing and accessories. Thus, a few portables appeared in leather or leather-covered cases; more common were portables in plastic that simulated leather—from alligator to snakeskin. RCA had an entire line of leather look-alikes.

Plastic, though, had won out decisively by 1950. Among the reasons for its triumph were improvements and cost reductions in the wonder material. Plastics could also be molded in myriad shapes and virtually any color; this cheap and infinitely flexible material was a designer's dream. Finally, introduction of the selenium rectifier—a pre-war semiconductor device—did away with the hot-running rectifier tube in three-way sets. This opened the door to making much smaller three-way portables because designers could now exploit lightweight (but heat-sensitive) colorful plastics. By 1948, almost all plastic portables (and many others as well) had selenium rectifiers.

Aside from the selenium rectifier, post-war portables boasted few technological innovations. In fact, the first portables on the market in 1946 were unchanged from 1942 models. Some companies had simply kept their stockpile of components and resumed production of the same set. For example, in 1946, Emerson's pre-war Model 432, the pocket radio, was sold again as Model 508.

A few radio companies claimed in ads that their portables embodied advanced technology developed during the war. In 1946, for example,

a

b

c

d

e

f

g

h

i

j

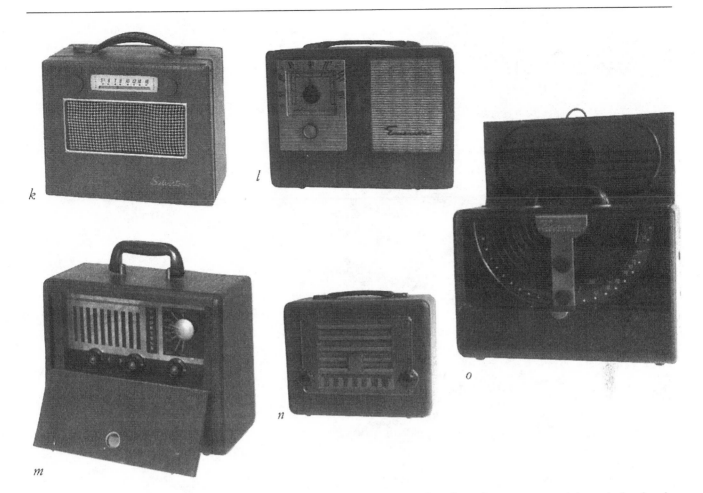

10.1 Portables in fabric-covered or painted wooden boxes (similar to the pre-war suitcase style) peter out during the late forties and early fifties. *a*, Trav-Ler 5020, 1946 (12″ wide) (rear view shows batteries in place); *b*, Admiral 6P32-6EI, 1946 (13″ wide); *c*, Stromberg-Carlson 1105, 1947 (13″ wide); *d*, Setchell-Carlson 447, 1948 (11-5/8″ wide); *e*, Automatic C-65X, 1948 (12-3/4″ wide); *f*, Sparton 6AM06, ca. 1947 (12-3/8″ wide); *g*, Artone R-546, 1947 (13-5/8″ wide); *h*, Trav-Ler 5028A, 1947 (8-3/4″ wide); *i*, Philco 53-658, 1953 (13-5/8″ wide); *j*, Fada P100, 1947 (12-3/4″ wide); *k*, Silvertone (Sears) 57DT226, 1951 (11-5/8″ wide); *l*, Emerson 754D, 1953 (12″ wide); *m*, Truetone (Western Auto) D3840, 1948 (13-3/8″ wide); *n*, Emerson 567, 1947 (9-5/8″ wide); *o*, Zenith 6G001ZYK, 1950 (15-1/8″ wide)

Motorola stated in *Holiday* that "because we originated the battle-famous 'Handie-talkie' Motorola portable radios are better than ever." The radios pictured, of course, contained pre-war electronics and were identical to pre-war models. Similarly, in advertising their Model 360 in the fall of 1947 (Fig. 10.2), Philco emphasized that the set had a special wartime circuit for greater sensitivity. Inspection of a 360 fails to disclose any novel war-derived circuitry. The radio does include an extra stage of amplification, so it would have been more sensitive than the run-of-the-mill, post-war portable. But Zenith sold many such super-sensitive portables before the war.

Even more outlandish claims were made in an article in *Radio and Television Retailing* on "Portables pile up profits." One purpose of the piece was to alert dealers to features of the post-war portables that could be highlighted in sales pitches, particularly "lessons learned during the war." The list of alleged war-derived improvements was, in fact, short and unconvincing. All were based on pre-war technologies.

One significant war-born innovation—the integrated circuit—finally did appear in pre-transistor portables in the late forties and fifties. The integrated circuit consisted of several resistors and capacitors on a ceramic wafer; the components were interconnected by silver paths "printed" on the ceramic wafer. (Resistors were also printed with a carbon solution.) The entire unit (except for protruding wires) was then plastic-coated. Made for commercial applications after the war by

10.2   An ad for the Philco 360 of 1947 that touted nonexistent, war-derived improvements in portable radios

Centralab, these components were called printed electronic circuits (to see one, look ahead to Fig. 12.6d). Perhaps because the name called attention to their unusual mode of manufacture rather than to the integrated product that resulted, in today's discussions of integrated circuits they are all but forgotten. But they were truly integrated circuits. Radio manufacturers used them, not to make better-performing radios, but to save space or to reduce labor costs (fewer connections to solder). A few portables contained them by the end of the forties, and in the fifties they were in common use.

Because the post-war portables were based essentially on pre-war technology, and played no better than the earlier sets, designers went to work creating a variety of gimmicks and distinctive features that could be touted in ads and in the showroom. Philco concealed the dials and station indicator of several models behind a wooden roll-top cover (Fig. 10.2), continuing the antique-desk look of its pre-war set. RCA's large Globetrotter portables (Fig. 10.3a) had aluminum cases with a sliding panel that served the same function as Philco's roll-top cover—whatever that was! The Motorola "Sporter" (Model 69L11) of 1948 and 1949 had the station dial in the carrying handle (Fig. 10.3b): "Easier to read . . . Easier to tune" boasted the ads. The Sporter also featured a woven fabric

a

b

c

d

case coated in clear plastic: "Its beauty is breathtaking." Maybe. Certainly it was a portable of unusual appearance—"a best buy at only . . . $49.95." General Electric's 145 looked like a small briefcase (Fig. 10.3d).

Even Zenith, a company whose reputation had rested on quality, got into gimmicks. Like Motorola, Zenith began to exploit previously unseen possibilities in the tuning indicator. In the "Flip Top" Holiday portables (Fig. 10.3e, f), the tuning indicator, which also contained the antenna, swung up and turned on the radio: "It's the 'tip-top' feature in radio! Swing the lid up—there's the dial actually *above* the set, its black numerals and 'glowing' red pointer giving tip-top tuning ease. And the famous Wavemagnet inside the lid tips up too, away from signal-killing metal parts."

e

f

g

10.3 Some post-war sets with "Distinctive Features." *a,* RCA Globetrotter 8BX6, 1948 (13″ wide); *b,* Motorola Sporter 68L11, 1948 (13″ wide); *c,* Emerson 505, 1946 (15″ wide); *d,* General Electric 145, 1949 (10-5/8″ wide); *e,* Zenith Flip-Top 4F40, 1949 (11-5/8″ wide); *f,* Zenith Flip-Top 5G41, 1950 (12″ wide); *g,* Zenith 6E40, 1948 (13-3/4″ wide); *h,* General Electric 260, 1947

h

In showmanship, the Tip-Top Holiday was a mere understudy to Zenith's "Universal Pop-Open" portable (Fig. 10.3g): "Just press the pop-open button and . . . everything happens at once! Doors pop open! Wavemagnet pops up! Set begins to play!" It even had a "lustrous modern finish in two-tone gray-beige," and was claimed to be "18 percent lighter." (Lighter than what? The ad did not say.) All this for only $54.70. Curiously, the ads never mentioned the one obvious advantage of this design: even when dead, the radio could still perform two of its three tricks.

Other radios, like the Emerson 505 (Fig. 10.3c), did nothing unusual, but they were large and, in the showroom, presumably impressive. General Electric's Model 260 had a rechargeable battery, doubtless a valuable sales feature (Fig. 10.3h).

For those seeking sheer novelty, Automatic offered the Tom Thumb Cameradio (Fig. 10.4), a combined portable radio and "fine reflex camera." Sold in 1948 for around $40, it was said to be "indispensable for trips, hikes, ball games, every outdoor and vacation activity." Needless to say, this was one portable radio that earned the right to look like a camera. Though it was on the market for several years, it is doubtful that many people found the Cameradio "indispensable."

Another memorable novelty portable—bordering on the truly bizarre—was the "Man from Mars Radio Hat" of 1947. This was a two-tube receiver built into a pith helmet. Actually, the tubes were not in, but on, the hat, protruding like horns; a vertically mounted loop antenna enhanced the overall ambience. The battery was kept in a pocket, and was connected to the hat with a rather obtrusive cable. This bit of portable whimsy sold for $7.95. Appropriately enough, the radio hat was made by American Merri-Lei, a company specializing in party goods. *Life* magazine and the *New Yorker* both ran articles on the Man from Mars Radio Hat; for some reason, it was judged to be newsworthy. *Life*'s piece, which included three pictures of people wearing the radio hat and mugging for the camera, summed up the phenomenon nicely: "set plays fine but looks ridiculous."

Portable radios were ideally suited for an increasing number of Americans who wished to partake of—and advertise—more affluent lifestyles. Like their pre-war predecessors, portables in the late forties and fifties continued to be regarded as a fine companion for trips, particularly to the beach (Fig. 10.5). Vacations by automobile were now nearly universal, and more than 80 percent of pleasure trips were by car. Visitation at recreation areas skyrocketed after the war. National parks, for example, went from 21 million visitors in 1941 to 46 million in 1953; by 1972, the number had climbed to 212 million. Not everyone brought along a portable radio, of course, but more people were taking part in the kinds of activities that the portable could accompany. At some beaches, portables could even be rented.

Sports of all sorts became much more popular because of television and the expansion of air travel (which made "national" leagues truly national). The avid sports fan wanting play-by-play everywhere turned with greater frequency to portable radios.

10.4   Automatic Tom Thumb Cameradio, 1948 (9-1/8″ high); *above,* radio side; *below,* camera side

10.5 A portable in use southern Florida, 1950

G-E Color-Styled Portables. Well earned gift–powerful portable in a striking maroon plastic case. Very light—only 8 lbs. with batteries. G-E Dynapower speaker. Plays beautifully on AC-DC or batteries. **$29⁹⁹**
Lowest priced 3-way G-E! Model 601.

10.6 General Electric ad takes aim at gift-givers, 1950

Home ownership surged dramatically as returning GIs married and established families. With new homes came home maintenance activities and an increase in do-it-yourself projects, many of which went better with radio. Outdoor pursuits, such as entertaining and gardening, also stimulated portable radio use. Lightweight three-way sets were especially favored for moving between rooms and between indoors and outdoors.

Hobbies of many kinds enjoyed increasing popularity. One way to while away leisure time was in assembling electronic kits. The electronic tinkerer could choose from a variety of portable radios sold with some assembly required. For someone with $15 or $20 as well as a screwdriver, pliers, and soldering iron, these sets could provide a few evenings of fun and, perhaps, frustration. All that was needed to follow the seemingly endless "step-by-step" instructions was patience, an ability to read fine print, and slender fingers.

Because they connoted happy times outdoors, portables remained the ideal gift for many occasions, including marriage, graduation, promotion, and retirement—not to mention Christmas, bar mitzvahs, and birthdays. By the late forties, much portable advertising was aimed at gift-givers, another distinct category of consumer (Fig. 10.6). The ads were effective. When young Americans went off to college, their parents sometimes gave them portable radios, especially three-way sets that could be used most of the time without draining batteries. Indeed, the portable radio entered the necessity category for the post-war college set.

The importance that the portable radio came to have as a gift is illustrated by one very unusual set, the Revere 400. Revere was a family-owned firm that made mainly movie cameras and tape recorders. Once in a while, though, the owners delighted in producing something different that was of high quality and "state-of-the-art." The Revere 400 portable radio of 1954 was one such product, the only stand-alone radio Revere ever made (Fig. 10.7). The radio looked like a binocular case and was fashioned from fine leather; when the snap was released and the lid raised, the radio turned on. The carrying strap doubled as an AC cord. According to Theodore Wickstrom, Revere's assistant chief engineer at the time, only one thousand of these radios were produced. The owners had no intention of competing in the cheap radio market; they simply wanted a unique product that, while potentially profitable (at $44.50 each), "could be given to good friends of the family and suppliers."

The most popular sets of the late forties and early fifties were "personal" portables in all-plastic cases about the size and shape of a lunch box. They were easy to carry, could be used in a variety of leisure activities, and—at $20 and up—were just right as gifts.

One of the first portables that helped to establish the plastic lunch box convention was the Emerson 560 (Fig. 10.8), a battery-only set introduced in the fall of 1947. Stylewise, it set the tone for sets to come. By 1950, all the major radio makers had their own plastic lunch box portables, mostly three-way sets, usually available in an assortment of colors (Fig. 10.9). Typical of these was the selection offered by Philco for its Model 631: Teal Green, Maroon, Caribbean Blue, and Swedish Red. In the early 1950s other styles were still being made, but the plastic lunch box reigned supreme.

In this golden age of the tube portable, an appreciable number of sets were advertised as "pocket portables." These ads—then and now—are confusing. Did "pocket portable" mean small enough to fit in the pocket

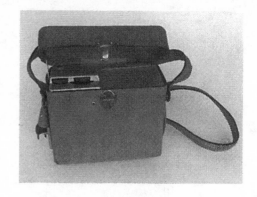

10.7   Revere 400, 1954 (7-1/2″ wide)

10.8   Emerson 560, one of the first plastic lunchbox portables, 1947

of my shirt? Or my wife's jeans? Or my dad's overcoat? Even a kangaroo would be puzzled. Obviously, radio companies used the "pocket" designation to advertise sets that were small compared to larger portables unable to slip into any known pocket. But, alas, pockets come in many sizes and so do portables. To reduce confusion I will try to specify pocket size.

Miniature tubes, of course, offered an unparalleled opportunity to shrink portables. With few exceptions, the pre-war camera-style radios, though relatively small, did not fully exploit the new tubes' potential. This was obvious to Kjell Gaarder, an RCA engineer who saw a few of the first BP10S, RCA's 1940 personal portable, while they awaited shipment to a trade show. In opening up the radio, he was surprised to find that the set contained quite a bit of empty space. If that were removed, he thought, perhaps the radio might fit a coat pocket. Gaarder took the set home and "adjusted" its case somewhat, shrinking it in several dimensions. Needless to say, his supervisors—who needed that radio in displayable condition—were not impressed with his ingenuity.

After the war, RCA and other companies set about making small portables, some suitable for coat pockets (Fig. 10.10). Early RCA sets included the Solitaire with "real gold-plate and luxurious catalin." RCA's most popular small set was the 54B (Fig. 10.10a), which was actually in production from late 1945 well into 1948. A camera-style radio with a metal and plastic case, it weighed three-and-a-half pounds, with batteries, and sold for around $26.00 (including batteries). The plastic lid of the 54B—molded to resemble alligator hide—popped open, turning on the set like a music box. The purchaser could select from brown, black, ivory, and maroon cases, though no alligators are found in the latter colors. This style was shamelessly copied by many manufacturers, who sold an assortment of nearly identical sets for $20 to $40.

In 1951 RCA introduced its model B411, arguably the best looking of the petite portables (Fig. 10.10f). Its chrome-colored grill would have been gaudy on a bigger radio, but was perfectly matched to this minimodel. Alas, beauty is sometimes fragile and transitory. Most of the surviving B411s are badly battered, the very thin Santay plastic case cracked. Durability was no longer of great concern to designers. To attract the gift-givers, a portable had to look good; protective covers were abandoned. If the set did not last, a new model could take its place next year.

Beneath its chrome smile, the B411 did have one noteworthy technological feature: a ferrite rod antenna. These antennas consist of a ceramic rod, containing iron, on which is wound a small antenna coil. Philco actually introduced this antenna in its Model 631 of 1950: "The Magnecor, an entirely new kind of aerial perfected by Philco research, stays permanently concealed inside the cabinet. There's no unsightly lid, loop, or rod to lift." These new antennas were uniformly adopted for transistor radios because they could be made much smaller than loops, but their use in portables began somewhat earlier.

Consumer magazines found few words of praise for the pocket portables. Overall poor performance and awful battery economy (seven to twenty-one cents per hour of operation) was a high price to pay for ex-

a

b

c

d

e

f

g

h

i

j

k

l

m

10.9 An assortment of plastic lunchbox portables, 1948–1954. *a,* Westinghouse H-185, 1948 (8-3/4″ wide); *b,* RCA 8BX5, 1948 (8-3/4″ high); *c,* Philco B652, 1953 (10″ wide); *d,* Motorola 58L11, 1948 (7-3/4″ wide); *e,* Firestone 4C13, 1949 (9-1/4″ wide); *f,* RCA 9BX56, 1949 (10-1/2″ wide); *g,* Motorola 5L2U, 1950 (7-3/4″ wide); *h,* Motorola 63L2, 1953 (9″ high); *i,* Emerson 646, 1950 (7″ high); *j,* DeWald E517, ca. 1951 (7″ high); *k,* Motorola 53L, 1953 (8″ high); *l,* Motorola 54L, 1954 (8″ high); *m,* Sears Silvertone 57DO217, 1951 (9″ high); *n,* RCA 2BX63, 1952 (7-1/2″ high); *o,* Zenith L406, 1953 (11-3/4″ wide); *p,* Capehart 1P55, 1954 (10-1/4″ wide); *q,* Philco 53-642, 1953 (6″ high); *r,* RCA BX57, 1950 (8-1/4″ high); *s,* Sentinel 319P, ca. 1948 (7-1/4″ high); *t,* CBS-Columbia 5220, 1954 (9-3/8″ wide)

treme portability. *Consumer Reports* in 1953 summed it up tersely: "Unless you are determined to buy the lightest and smallest portable you can, there appears to be no good reason for buying a 'battery-only' set."

Despite a lack of praise and performance that fell far short of larger sets, the coat-pocket portables—even battery-only sets—were bought in sufficient numbers to keep the genre alive. A 1947 article in *House and Garden* judged these small portables to be appropriate for women. (Almost certainly many were given as gifts—perhaps to women—with no thought to battery economy.) Many of the large manufacturers continued to include one or two pocket portables in their radio lines in the early 1950s. There was obviously a market for these tiny portables, but probably it was not a large market; otherwise, all radio makers would have offered a wide choice of models.

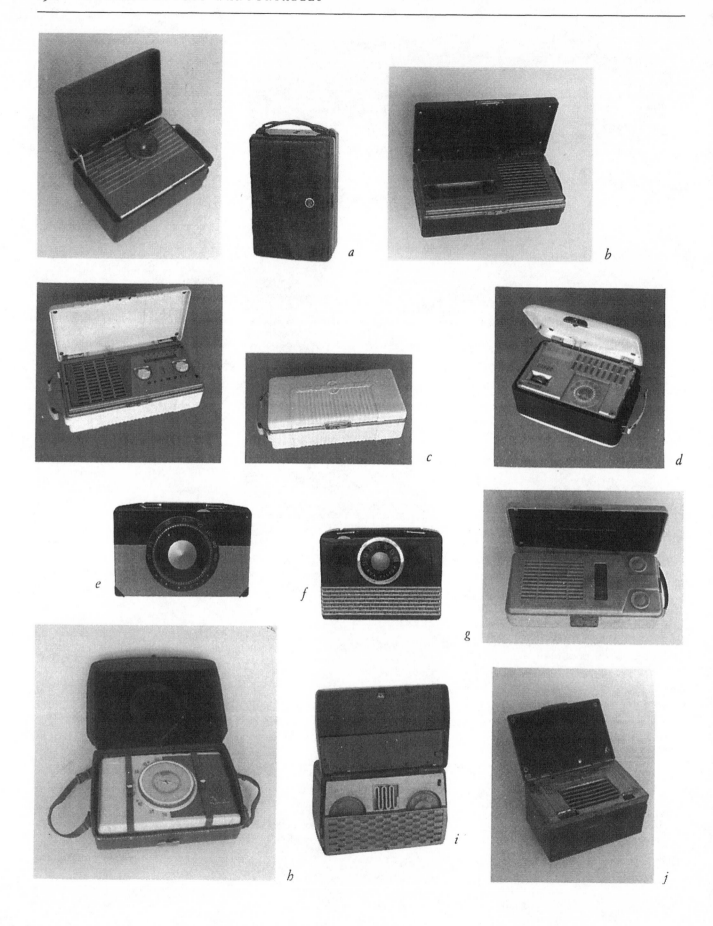

a

b

c

d

e

f

g

h

i

j

k

l

m

10.10 Some small portables of the post-war era. *a,* RCA 54B2, 1946 (6-1/8" wide); *b,* Emerson 558, 1947 (8-1/2" wide); *c,* Olympic 449, ca. 1948 (8-3/4" wide); *d,* Garod 4A, 1947 (6-3/4 wide); *e,* RCA 2B401, 1952 (8-3/4" wide); *f,* RCA B411, 1951 (7-7/16" wide); *g,* Emerson 640, 1950 (9-1/2" wide); *h,* Sylvania 454, 1954 (9" wide); *i,* Motorola 52M, 1952 (7-1/2" wide); *j,* Air King 494, ca. 1948 (6" wide); *k,* Sears Silvertone 8260, 1946 (8-1/2" wide); *l,* Sentinel Treasure Chest, 1947 (8-1/4" wide); *m,* Emerson 508, 1946 (8-1/2" wide)

Surprisingly, even smaller portable radios were possible, though not with miniature tubes. Indeed, a five-tube portable, petite enough to slide into a shirt pocket, was produced commercially (see Chapter 12).

On the other size extreme were the Zenith Trans-Oceanic and its clones. Though redesigned, the Trans-Oceanic actually grew after the war (Fig. 10.11b). In fact, it was much bigger than was necessary merely to accommodate its electronic guts. Two factors contributed to this radio's obesity. First was the need for a big baffle for the speaker; a radio of this kind (and cost) needed to sound as good as a decent table model. Second, and most important, was the set's enormous battery pack. The huge battery allowed the set—which in one version had seven tubes plus rectifier—to operate very economically, around two cents per hour (about the best of any tube portable), a figure that would not be appreciably bettered until the advent of the transistor radio. The Trans-Oceanic was redesigned in 1951, but it got no smaller (Fig. 10.11a).

The Zenith Trans-Oceanic series was one of the most successful portables of all time, despite its cost ($87.50 and up). With its short-wave bands, the Trans-Oceanics provided access to the world and the appearance of worldliness to its owner. It was used as a prop in fashion layouts, and advertised in upscale magazines, perpetuating its image as a necessity for the socially with-it. An ad in *Holiday* from 1948 showed a middle-aged man smoking a pipe, listening to his Trans-Oceanic poolside. To underscore this image of success, the ad says: "Of course, it's a Zenith . . . The Aristocrat of Portable Radios. The Trans-Oceanic—its owner list reads like 'Who's Who in the World.' The choice of statesmen, executives, leaders in *all* walks of life." The meaning of the Trans-Oceanic had become quite clear. With its upscale image, the set would become especially desirable among Americans of the middle class, whose social aspirations and material cravings then had no limits. Like all successful products, the Trans-Oceanic had a host of imitators, mostly in the fifties (Fig. 10.11c, d), but none had the impact of the original.

Although the Trans-Oceanic and its clones had four to seven bands, none included FM. In fact, no major radio company made a tube portable with FM. There were two principal reasons why FM did not get a foothold in portables until the late fifties. The first was that FM has a limited reception distance, rapidly fading after about twenty-five miles. Thus, on outings in the country, no FM stations would be heard (this is still true today). Second, manufacturers apparently thought that classical music mavens would be unwilling to suffer the inferior sound quality of a portable radio, most of which had rather small speakers (4" or less). As it became obvious in the early sixties that portables were often being used around the home within range of FM stations, U.S. radio makers finally began to issue portables with FM. However, they did not become popular until the late sixties.

From 1946 to 1954, portable radio sales exceeded 1 million sets annually. In two years, 1947 and 1948, sales of 2.5 million were registered. When the first transistor radio—a portable—arrived in December of 1954, it joined more than fifteen million portable tube sets bought by Americans in the post-war years. Clearly, the portable radio had found a

a

b

c

d

10.11   Several multi-band portables of the post-war era. *a,* Zenith Trans-Oceanic H500, 1951 (17″ wide); *b,* Zenith Trans-Oceanic 8G005YT, 1948 (17″ wide); *c,* Hallicrafters TW-500, 1954 (15″ wide); *d,* RCA 3BX671, 1953 (17-1/2″ wide)

permanent place in the lives of many Americans, most of whom were enjoying a lifestyle once reserved for the wealthy. Portables were no longer luxuries; in the late forties and early fifties, a nice portable could be bought by a factory worker with just two to three days' wages.

By the end of the fifties, tens of millions of families had portables, many of them shirt-pocket size. Let us now turn to the factors—social and technological—that made it possible to realize this ancient cultural imperative, a radio of near-ultimate portability.

11

Child's Play

THE LATE FORTIES BUYING BINGE AFFECTED EVERY
stratum of American society. Not only did the wealthy get their first
"home entertainment center," but a poor family might have bought its
first electric refrigerator. The dimensions of the consumer society en-
larged in other ways as well. During the post-war years, and especially in
the fifties, the purchasing power of young Americans (children and teenagers) and their influence on parents' purchases became explicitly recog-
nized by manufacturers and Madison Avenue. And so, as new products
entered the kitchens and living rooms of ordinary homes, still other new
things—some of them electronic—took up residence in back bedrooms,
as youth took up new kinds of play.

Children have always had a world of their own tiny things to play with,
especially scaled-down adult objects. In Bellamy's time, for example,
boys and girls played with diminutive rubber animals, warships, toy soldiers, cannons and guns, wagons, musical instruments, china tea sets,
dolls, kitchen utensils, and even sewing machines. With toys of this sort
children could mimic the activities and roles that they would eventually
assume as adults. Sometimes children's toys pointed to the future: things
that no one—large or small—yet owned, like rocket ships and ray guns,
and pocket radios.

As the electrical age began to assert itself in the affairs of adults, electrical toys became fair game for children. Among the earliest was the
Boston Motor and Battery Outfit, sold in the 1890s. It came with a four-
inch fan—too small to do much—and cost $1.25, not a trifling sum.
About the same time one could buy a $2.25 Novelty Medical Coil and
Battery, which children used to test each other's willingness to endure
electrical shocks. In the first decade of the century, battery-operated electric trains began rolling, though in quite small numbers; not until the
twenties would electric trains that could be plugged into AC become fix-
tures in many homes. Erector sets with electric motors also achieved
popularity by the twenties.

Although electric trains were impressive toys, most miniature things
were so reduced in size and simplified in design that they became com-

pletely nonfunctional. Solid wooden trucks that did not roll and stoves that produced no heat stimulated young imaginations, but they were "only toys." In the adult world, labeling something a "toy" was not a compliment to clever design; rather, it indicated that a product had been fatally down-sized and was incapable of performing well (or at all). Fear of the "toy" label had a chilling effect on established companies, concerned with maintaining their good reputations. Such a company would not deliberately down-size its products to the point where consumers might treat them as toys. This concern was especially common in the electronics industry, where too much miniaturization or simplification could easily force a product into the toy category (as International discovered in 1933 with its Kadette Junior). Thus, beginning in the late twenties major radio companies left the manufacture of toy electronic products to the entrepreneurial fringe and to established toy makers so that their image—as a supplier of serious products—would remain pure.

The possession of home radios by children goes back, of course, to the amateur era in the first decade of the century. But most children—then and now—were not electrical experimenters; they needed a ready-to-use radio with minimal (if any) assembly. In the twenties, a variety of tiny companies continued to make crystal sets—some pocket size (Chapters 4 and 6), many of which doubtless ended up in the hands of youth. The advent of the midgets in the thirties also opened up opportunities for children's radios. Midgets were heavily promoted as a second and third set to more affluent families, and a few midgets certainly found their way into back bedrooms. After the war, midgets in large numbers were eagerly cast off by middle-class adults and just as eagerly grabbed up by their children—a perfect match.

In the late forties, tiny table models and small portables were advertised as ideal gifts for children, but they were not pushed as toys per se. Indeed, most were superheterodynes that met reasonable standards of sensitivity and selectivity, even if their sound was somewhat less than high fidelity. On the other hand, the radios aimed only at children were, as crystal sets, true toys.

In the decade following World War II, the crystal set became the first radio that most boys bought or were given, and they can still be purchased today. Their enduring fascination probably lies in their simplicity; and, of course, their low price.

Because they have few parts, crystal sets also lend themselves to extreme miniaturization. Recall that tinkerers in the teens and twenties had built crystal sets in match boxes and watch cases and on pinkie rings (Chapter 6). As commercial products, miniature crystal sets would, in the thirties, forties, and fifties, spread the cultural imperative of pocket radios to millions of children who were not among the radio experimenters. The Sunday comics also figured importantly in popularizing radios of extreme portability.

Probably the most famous miniature radio of all time appeared in the comic strip *Dick Tracy*. The detective hero, whose name twice reveals his profession, had a two-way radio in the guise of a wristwatch. Although the comic strip began in 1931, Dick did not get his radio watch until the

11.1 Dick Tracy radio box, Da-Myco Products Co., ca. 1949

mid-forties. In fact, Chester Gould, creator of the comic strip, majored in business administration at college and was not much of a techno-mancer; Dick Tracy rarely relied on technical aids during his capers. The wristwatch radio originally was envisioned as a two-way radio-television, but Gould dropped the television capability because, since the new medium "was still in its infancy," a television that small would not have been believable.

A wristwatch radio was more than merely believable by the end of World War II; spy radios had probably been built that size already. The wristwatch radio itself immediately became a cultural imperative among experimenters, who would build them with a variety of technologies. In addition, though Dick Tracy's diminutive timepiece radio was not used for entertainment, it became for decades the standard against which small portable radios were measured in the popular media.

Not surprisingly, the first commercially made Dick Tracy wrist radios were crystal sets. One early example was made around 1949 by Da-Myco Products Company of New York City; the box (Fig. 11.1) was much more memorable than the radio it contained. There were no exaggerated claims for this set; the instructions clearly admitted that "THE DICK TRACY WRIST-RADIO is only a toy radio after all." Though lightweight and small, the set still needed an external antenna and ground. Nonetheless, the message that children got from this tiny radio was very clear: as they grew up, perhaps progress would bring them fully functional wrist radios. The Dick Tracy radio and other radio toys introduced young Americans to the expectation that radios would become ever more portable.

Though the miniature radios of children helped to perpetuate the cultural imperative of ultimate portability, they also reinforced the belief, held by many adults in the major radio companies, that miniature radios were toys. Companies on the fringe, though, appreciated that the youth market for such radios was potentially profitable; they could plunge in without worry that they might tarnish their image—they had no image. One such company, Western Manufacturing, achieved remarkable success selling small crystal sets. Moreover, some of them had certain design features that anticipated—and very likely influenced—the look of later transistor shirt-pocket portables.

Western Manufacturing Company was located in Kearney, Nebraska, home of Paul Spurgeon Beshore. While still in his teens, Beshore began to build and sell small crystal sets. His first mini-radios—ten thousand of them—were housed in pill boxes and sold by mail for a buck or two.

The mail-order sale of crystal sets as well as plans for other items (including a motor scooter eventually produced by Cushman) furnished in the Depression the foundation of Western Manufacturing Company. In this venture Paul Beshore was joined by his brothers Woodrow and Charles. Paul, the creative force behind the products, attended college for only one year, but he was an insatiable reader who kept abreast of many fields. The company's approach to manufacturing reflected Paul's style, which was to maintain complete control over everything, including obtaining the galena for crystals in Colorado, winding the coils (originally on wooden blocks), and making the cabinets.

To keep costs low, crystal set makers before Western simply mounted the few components on a piece of wood or bakelite in plain view; the cheapest crystal sets seldom had a housing. Actually, ready access to the radio's innards was necessary because of the near-constant need to fiddle with the positioning of the cat's whisker on the crystal. A few firms solved this problem with ingenious mechanical controls that allowed adjustments from the radio's front panel. However, this solution upped the price of the set. In keeping with his desire to minimize costs, Beshore employed another solution, which was to fix a cat's whisker permanently to the crystal and enclose both in a small plastic tube with contacts at each end. In this way, the always-functioning crystal could be out of reach (and sight). These no-adjustment crystal sets, housed in cute cases, were advertised in hobbyist magazines as "pocket radios" (Fig. 11.2).

During the Depression other crystal set makers sold a variety of pocket sets (Fig. 11.3), though none lasted very long.

Around 1940 Beshore created for his crystal sets a molded plastic case that looked like a miniaturized table radio (Fig. 11.4). The earphone as well as antenna and ground wires were plugged into the sides of the radio. The micro-table-model design remained in production for many years, undergoing minor changes—sometimes annually.

The micro-table-model was very dainty, weighing only six ounces, and it nestled easily in the hand. The set did little but it also cost little—$2.99 (postpaid). It carried the brand names "Midget" and "Tinymite" and perhaps others. Though lightweight and petite, the micro-table-model still had to be hooked up to an antenna and ground, reducing its portability.

Like many radio makers during the Second World War, Western dedicated its plant to military production. After the war, Western resumed making crystal radios and expanded its operations by selling assorted items from many companies through mail-order catalogs, offering everything from roller skates to the Tom Thumb Cameradio.

Other companies, perhaps impressed with the longevity of Western's charming crystal set, offered their own versions. For example, in 1947 appeared the Revell Radaradio, a clear clone manufactured by a toy company. The crystal in this radio, as in most post-war crystal sets, was a germanium diode, a product of radar research in World War II—hence the name, "Radaradio." The same table-model cabinet, by the way, housed the Modernair 500, a one-tube battery radio. Both sets were made in Los Angeles, where small table radios got their start in the Depression.

Immediately after the war, Western introduced a radically new style of crystal set that would sell well (at $3.99 to $6.99) for more than a decade and a half (Fig. 11.5). These sets embodied unique design features that enhanced portability. The plastic case, molded in two pieces held together by screws from the back, was completely redesigned and shrunk even more; the radio easily fit in the palm of the hand or in a shirt pocket. To improve portability even further, Western built the earphone into the case. In addition, short wires for the antenna and ground emerged from two holes in the side of the case and had clips at their ends.

11.2   Ads for Depression-era "Pocket Radios" made by Western Manufacturing Company

11.3 Ads for Depression-era "Pocket Radios" made by two Chicago companies

These clips and the built-in earphone gave Western's pocket radio unprecedented portability for a crystal set. To listen, one simply pressed it to an ear and went in search of an antenna, attaching the clip to various metal objects in the home—bedspring, large metal lamp, and so forth. If furniture did not suffice, the coil of wire that sometimes accompanied the radio could be strung outdoors to improve reception. An appropriate ground was also needed; for this purpose, the second wire could be clipped onto a radiator or pipe. In practice, at only several feet long the wires were too short: sometimes a suitable antenna and a satisfactory ground were in different rooms.

The first of Western's post-war pocket radios (Fig. 11.5, top left and right) was marketed under the brand name "Pa-Kette" by the Pa-Kette Electric Corporation, Dept. PS-2, Kearney, Nebraska. This was one of many instant brands and instant companies that Western employed as part of its marketing strategy. "Dept. PS-2" was a code for the magazine and the month in which the ad ran (*Popular Science,* February); this was a means to monitor which ads in which magazines were effective. Later sets were marketed under different brand names, including Tinytone (Fig. 11.5, bottom left and right), Mitey Pocket Radio, and Pee Wee, some of which were made in different styles and advertised as new models.

Though not portable in every sense, Western's crystal sets were a unique and audacious design—probably inspired by pocket light meters—that attracted numerous customers. In the post-war period, buyers of the radios most likely were children or parents who purchased them as gifts. In many ads, these "pocket radios" were claimed to be the "ideal gift for children." By then-current standards, the crystal set was a toy; yet, even today, the design is nifty and appealing. Surprisingly, the Pa-Kette pocket portable was reviewed as a serious radio by *Consumers' Research Bulletin* in 1947. Needless to say, they did not like what they saw, complaining about lack of sensitivity and selectivity, and poor tone quality. The article concluded with a statement of the obvious, "This radio is definitely in the novelty or toy class . . . purchasing one would be a waste of both time and postage for most people." However, the children who sent away for the Pa-Kette and others of its genre knew nothing about technical standards; what they wanted was a tiny portable radio, and that is approximately what they got. Needless to say, most kids got bored quickly and were eager to move up to something better. They were getting practice at becoming good consumers.

From 1946 onward, Western Manufacturing's pocket radios were advertised relentlessly in *Popular Science, Popular Mechanics,* and other magazines. At the time, *Popular Science* alone had over a million subscribers. Even if most readers did not succumb and buy the comely set, the idea became fixed, month after month, that a palm-size, pocket portable was a desirable product. This cultural imperative was passed on to a new generation—of potential buyers and future radio engineers. In spite of the fact that they were toys, Western Manufacturing's crystal portables helped establish design conventions for very tiny pocket radios. They may not have been engineering wonders or portable in all respects, but they were seen repeatedly by radio people in their own magazines, such as *Radio & Television News,* and so would have had a subtle influence. When transistor shirt-pocket portables appeared in the mid-fifties, they would resemble Paul Beshore's mini-radios.

The most significant electronic toy of the post-war era was the phonograph. Indeed, many middle-class children received "record players" as gifts in the late forties and early fifties long before their parents purchased a hi-fi phonograph. Let us see how this curious state of affairs came about by picking up the story of the phonograph after the advent of commercial broadcasting.

When radios began to sell in large numbers in the late twenties, the phonograph industry fell on hard times. The established phonograph companies were very slow to convert to electronic systems, which then were expensive. Not surprisingly, in the Depression most people got rid of their old phonographs and tuned in to radio; the latter was more entertaining, sounded much better, and it was free.

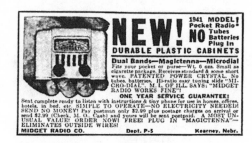

Records and record players limped along during the Depression with a few very expensive electronic models available. Electronic phonographs reproduced what was on the record quite faithfully; the problem was that, after a few plays, static and scratches became an ever larger part of the program. Caruso and other great artists had to suffer many indignities. To manage the scratchiness problem, manufacturers usually put large speakers—sometimes fifteen inches across—in their expensive console radio-record players. Because such speakers could not handle high frequencies, the scratches were mostly eliminated, but so was some of the music. By the end of the decade, about the same time that portable radios revived, the buying public began to show interest in down-sized radio-phonograph combinations, which could be bought for as little as $29.95—but an automatic changer was more. In the view of many company executives, the record industry seemed poised for expansion.

11.4   Examples of Western Manufacturing's "Midget" pocket radios with plastic cases

After the war, RCA and CBS, which both made records, invested in new record technologies. Both companies turned to vinyl plastic (Vinylite) instead of shellac, which made lighter, less brittle records; the new material also reduced scratchiness. Both companies also employed smaller grooves and placed them closer together, thus decreasing the amount of room required for a given piece of music. However, there the similarities ended. RCA made a small record that could play one popular song on a side; CBS made a large record that was well suited to classical music and shows. And so, in the late forties were born the two types of records—one played at 45 rpm, the other at 33 rpm—that would remain standards to this day.

Almost immediately, CBS and RCA offered record changers that could be played through the buyer's existing radio-phonograph system. Very soon, however, many companies began to sell small, relatively inexpensive ($25 and up) record players that could handle the new 33s and 45s as well as 78s. Many of these record players, which usually had a feeble amplifier and four-inch speaker—adequate for playing popular music or children's records—were built into small, child-size suitcases. In a pattern that has become familiar, these AC-only record players were advertised as "portables."

11.5 Some of Western Manufacturing's post-war pocket crystal sets (3" high, 2-1/8" wide). *Top left and right,* Pa-Kette (late forties); *bottom left and right,* Tinytone (late fifties and early sixties)

Because the portable record players were within reach of a middle-class budget and the number of children's records had increased dramatically with the coming of unbreakable Vinylite, parents saw this artifact as a most wonderful gift. Children would be able to while away countless hours without needing close supervision. They would also gain confidence in their ability to master an electrical product that made music.

Record players available to adults included expensive TV-radio-phonograph combinations and equally expensive component hi-fi systems. Stand-alone phonographs that played well by adult standards could also be bought for about $50 and up. In many middle- and working-class homes, getting a TV had a much higher priority than a phonograph. And, once the TV entered—and dominated—the living room, the need for a phonograph did not seem so pressing. Not until the sixties, with the advent of inexpensive stereos, would most middle- and working-class families have a phonograph that adults had bought for themselves.

Prior to the advent of children's record players in the late forties, the largest market for popular music singles (78s, of course) was the 400,000 jukeboxes in bars, dance halls, cafes, and the once-ubiquitous corner drugstore. A "hit" record was one that got lots of jukebox play. After the war, popular music programs on the radio, staffed by "disc jockeys," began to stimulate huge sales of records to young Americans, mostly teenage girls. This trend, of course, was facilitated by the spread of the "portable" record player. By the early fifties, the teenager record market—mostly for 45 singles—was very well developed. Articles in trade magazines furnished advice to dealers on how to cultivate teenage consumers.

With their allowances or money earned from odd jobs, teenagers could buy their favorite records and play them in the privacy of their own bedrooms. When they were tired of playing records, they could listen to popular music on the radio—the same stations that Mom played in the kitchen. While listening to music, teenagers could read magazines, like *Seventeen*, that provided unremitting guidance on how to become perfect consumers in a consumer society.

When the musical preferences of parents and children began to conflict in the mid-fifties, shirt-pocket portable radios would help to provide a resolution. Let us turn to the technologies—subminiature tubes and transistors—that made it possible to build shirt-pocket superheterodynes, a cultural imperative realized.

12

Revolution in Miniature
The Shirt-Pocket Portable

IN 1940 AND 1941, WHILE AMERICANS WERE SNAPPING up portables at an unprecedented pace, England was already at war, taking punishment from Nazi aerial bombardments. One response to Germany's air war was a radarlike transceiver so tiny (about the size of a small Campbell's soup can) it could be installed in an anti-aircraft round and shot from mortars. When the radio device determined that the target—a rocket, aircraft, or buzz bomb—was within range, it detonated the flying bomb. A gunner could actually miss the target but still score a hit. Called the proximity fuze and perfected by the U.S. National Bureau of Standards in 1943, this device greatly improved the effectiveness of aerial defenses and helped England down a significant number of weapons hurled at her through the air.

At the heart of the proximity fuze were the ancestors of Centralab's integrated circuits and a set of new tubes—subminiatures—produced by U.S. manufacturers Sylvania and Raytheon. These tubes were astonishingly small, a typical example measuring 1.6" long and 0.3" by 0.4" in cross section. By the end of the war, still smaller subminiatures had been made, one type being seven-eighths of an inch long. The person most responsible for this tube, which was actually developed before the war, was Norman B. Krim, a Raytheon engineer. In the late thirties Raytheon mainly made tubes and had considerable expertise in battery tubes. Having learned that the British company Hivac, with help from a former Raytheon employee, was producing exceedingly small, low-drain tubes for hearing aids, Krim did some market research in the United States. His poll of hearing-aid manufacturers in early 1938 revealed many potential customers for subminiature tubes. Krim proposed the hearing aid tube project—which was estimated to cost $25,000—to Raytheon's president, Laurence Marshall. Marshall asked Krim if he would quit in the event that the tubes failed to earn back the development costs. Krim said yes and the work began.

Making such tiny tubes presented problem after problem, from getting the composition of the filament coating just right to creating a sufficient vacuum. In March of 1939, after months of trial-and-error, Krim and his colleagues succeeded. Hearing-aid companies immediately wanted samples of the new amplifier tube (CK501X), and by the outbreak of the war many of them were selling hearing aids with the new tiny tubes. The tubes turned a profit quickly, and Krim kept his job.

As a result of events beyond the control of Raytheon, in the early forties military uses of subminiature tubes became paramount. For these peculiar applications, the tubes were ruggedized and new varieties made. And subminiatures underwent the transition in name from "hearing-aid" tubes to "proximity-fuze" tubes. Defense work was good for Raytheon, a company that had only $3 million in sales before the war; at its end annual income had risen to $173 million. In early 1945, as the conflict drew to a close, Raytheon executives (which now included Norman Krim) held a strategy session at Laurence Marshall's home to discuss various projects for maintaining the company's prosperity during peace time. Many ambitious high-tech projects were suggested, including microwave ovens (using Raytheon's radar tubes), televisions, microwave communication systems, and assorted mobile radios. When it was Krim's turn to propose a product, he picked something a bit more prosaic that could, nonetheless, effectively exploit subminiature tubes. His choice? A shirt-pocket portable radio.

It is not difficult to surmise where this idea came from. Krim had been a radio amateur and as a child during the twenties had devoured Gernsback's radio magazines and built sundry apparatus. Perhaps he had seen the WestingMouse portable with its subminiaturelike tubes, or maybe he had responded to Gernsback's incessant editorials calling for ever-smaller portables. One way or another, Krim had become a member of the shirt-pocket radio constituency; unlike other members, though, he was in a position to invest a sizable sum of money turning that cultural imperative into commercial reality. Laurence Marshall again gave Krim the go-ahead. (It would cost Raytheon about $50,000 to design their Lilliputian listening device. This seems like a lot of money for the time, but Raytheon as a major defense contractor was subject to a 90 percent excess profits tax—$50,000 was really only $5,000.)

In mid-1945, Raytheon bought the Belmont Radio Corporation, a Chicago firm whose bread-and-butter had been the making of "private label" sets for Montgomery Ward, Western Auto, and other mass merchandisers. Raytheon's new consumer products would be built by Belmont; if demand for the shirt-pocket radio was great, perhaps new distributors might be eager to adopt the entire Belmont line.

Krim assigned a Raytheon engineer, Niles Gowell, the task of designing the world's first commercial shirt-pocket radio. This was not to be a toy radio, but a superheterodyne. Working closely with Lemuel Temple, a battery expert, Gowell created a sophisticated five-tube radio small enough to slide easily into a shirt pocket (Fig. 12.1); it was only five-eights of an inch thick. Although a few parts for the audio section were the same as in hearing aids, suppliers had to craft some new miniature components for the radio frequency and intermediate frequency stages. Also, new types of subminiature tubes were created for this set (Fig. 12.2). The A battery consisted of two penlite cells, and a small hearing-

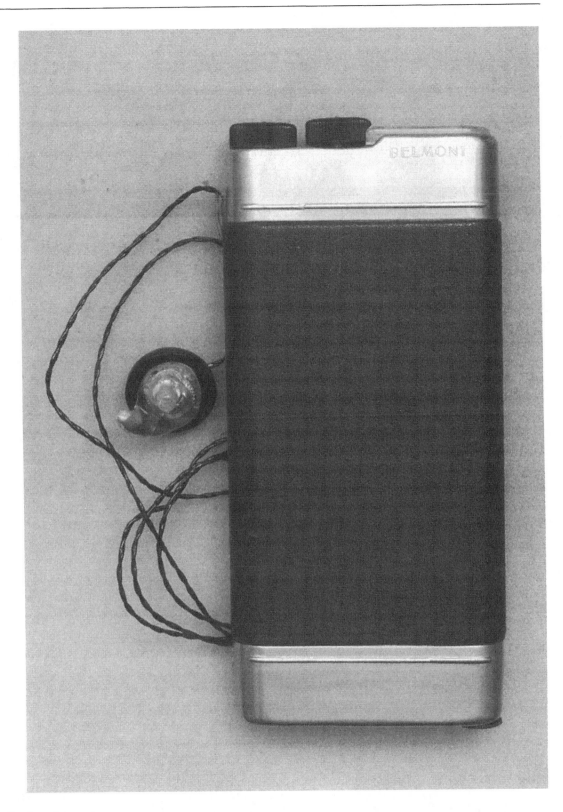

12.1   The Belmont Boulevard (5P113), 1945
(actual size)

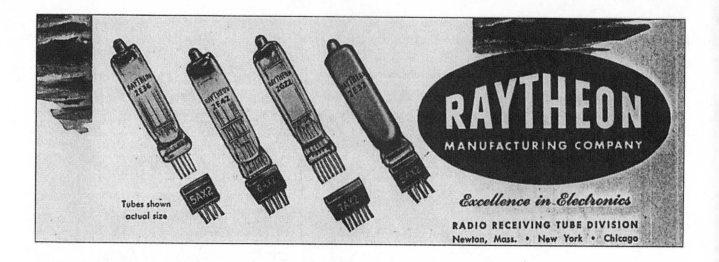

aid battery supplied the B voltage (22.5 v). The listener used an earphone, not a speaker, just like a hearing aid.

The general appearance and intended functions of the Belmont Boulevard seemingly had been influenced by earlier designs for pocket "sportsets" published in hobbyist magazines. However, the set's specific dimensions and the configuration of its controls closely resembled a hearing aid. The Belmont Boulevard's case also vaguely suggested a tobacco tin, which at that time was still a common sight in a man's pocket. (Remember Prince Albert in a can?) Though compact and elegant, the Belmont Boulevard overall was a somewhat masculine artifact.

When presented with the shirt-pocket radio, Belmont people were not impressed; after all, *they* knew radio and radio retailing. In their considered opinion the set would have few buyers because—despite its circuitry—it was a novelty item. With Raytheon firmly in control, however, Belmont agreed to make and market this minuscule radio.

The set entered production in late 1945, and its birth was announced in December to the American public in a full-page ad in *Life* magazine. A signed painting pictured the Belmont Boulevard, actual size, nestled between a set of keys and gloves. A "war-born" subminiature tube was shown, which helped to emphasize the radio's technical virtuosity. The ad was clearly directed at the trendsetters, and suggested how the pocket radio might be used:

> With the Belmont Boulevard in your pocket or purse, you can tune in on drama, news, and music as you walk along the street. You can attend sports events and hear a radio description of the thrilling action before your eyes. You can be a bedtime listener without disturbing the sleep of others. You can catch favorite programs you would otherwise miss.

Blurbs on the Belmont Boulevard were published in general-circulation, hobbyist, and electronic magazines, including *House and Garden, Popular Mechanics, Popular Science,* and *Electronic Industries.* All in all, the Belmont Boulevard was presented to the public in a very

12.2 Subminiature tube types used in the Belmont Boulevard (shown actual size)

favorable light. Surely the demand for subminiature tubes stimulated by sales of this set would add some volume to Raytheon's hearing-aid-tube assembly lines. That, at least, was the view at Raytheon.

As Belmont had predicted, however, customers did not swarm around stores to await the arrival of the first shirt-pocket portable. No one was trampled on the way to the sales counter. In fact, there was hardly any customer reaction at all: *total* sales reached, at most, 5,000. Quietly but quickly the proto-Walkman died. Americans of the immediate post-war period apparently were not ready for a shirt-pocket portable, at least not one that cost around $30 to $65 (sources disagree on the exact price).

Despite the minor setback, Raytheon continued to tout the virtues of its subminiature tubes for building an assortment of civilian products, including pocket radios. Sylvania was in a similar position as Raytheon, seeking new markets for its subminiatures. Failure to find such markets meant, of course, idle production machinery and layoffs. Both companies ran full-page ads in the *Proceedings of the Institute of Radio Engineers,* alerting engineers to the new products now made possible by their tiny tubes. In one especially visionary promotion in 1946, Sylvania stated that their tubes would lead to small "table-top" televisions (when only a few Americans had a television of any size). Raytheon, too, said the day had arrived when policemen and firefighters could routinely use pocket radios. In the spring of 1949, Sylvania ran a full-page ad showing a pocket portable—embraced by a comely model—having four subminiature tubes. The ad asks: "In a radio set, how small can you get?" Though Sylvania apparently had the answer, major radio companies, it seems, were not asking the same question. The handsome radio pictured in this ad was doubtless a mock-up, never put into production.

Raytheon was not secretive about its designs for the pocket radios themselves. Far from it. Demonstration circuits, sets of subminiature tubes, and technical information were made available to any interested company. A few firms, like Sentinel, actually made prototype shirt-pocket radios to gauge public reaction, but none entered production. Indeed, after the Belmont Boulevard fiasco, no *major* radio maker used these tubes in a shirt-pocket portable, despite Raytheon's optimistic claim in 1946 that "all progressive radio manufacturers anticipate the tremendous possibilities inherent in the small pocket receiver built around Raytheon standard subminiature tubes."

Interest among electronic hobbyists, however, was high. After all, subminiature tubes provided experimenters with a highly appropriate technology that they could use themselves in realizing—and perpetuating—the cultural imperative of a shirt-pocket portable. Not surprisingly, plans for such sets preceded the Belmont Boulevard. Especially noteworthy was L. M. De Zettel's three-tube set described in *Popular Mechanics* in December 1941. In electronics and hobbyist magazines of the late forties and early fifties, designs for pocket radios and other devices with subminiature tubes were especially common. For example, in 1948 *Popular Science* presented a pocket radio that could "double as a hearing aid"; *Popular Mechanics* also had plans in 1949 for a pocket set using subminiature tubes.

The high post-war profile of subminiature tubes and the potential for using them in shirt-pocket portables did not go entirely unnoticed in the commercial world. A small entrepreneurial fringe sprang up to exploit the possibilities. Perhaps the earliest radio to appear was the Pocket-Mite, sold in kit form for $6.99 (Fig. 12.3). Micro-Electronic Products of Peru, Indiana, marketed a three-tube shirt-pocket set in late 1948 and 1949 for $19.75 (or $15.75 as a kit); it weighed only 5.5 ounces, complete. Like the Belmont Boulevard, it played through earphone only. In 1950 Micro-Electronics offered the first shirt-pocket FM receiver for $37.50.

The Privat-ear (Fig. 12.4) was the most stylish and popular set of this genre (at around $20 to $30), enjoying sales from 1951 to about 1954. A two-tube set, the Privat-ear employed a reflex circuit in which the same tubes functioned in both radio- and audio-frequency stages (thus providing four-tube performance). Like the others, it had an ear plug and telescoping antenna, the latter making use in the pocket somewhat awkward. Gernsback's *Radio-Electronics* wrote up the Privat-ear in 1951; it was said to "provide sensitive reception with adequate volume"—even "pulling in stations from as much as 50 miles away."

Although a modest success, the Privat-ear and other shirt-pocket radios were probably bought mainly by electronics buffs and novelty seekers. They certainly did not reach a mass market or attract attention in mass-circulation magazines. What is more, that neither *Consumer Reports* nor *Consumer Research Bulletin* reviewed these radios indicates their status in the late forties and early fifties as novelties or toys. Though a cultural imperative for decades, the shirt-pocket portable (with subminiature tubes) did not really catch on in the marketplace. (However, some of the more specialized communications uses of shirt-pocket radios with subminiature tubes did come to pass in the late forties and early fifties, such as pagers and citizen's band transceivers.)

Norman B. Krim suggests several factors to account for the Belmont Boulevard's failure to attract buyers. Most important was the set's power consumption. Though low-drain, the filaments of the five subminiature tubes still used the same power as three or four miniatures; as a result the life of the A battery—two penlite cells—was mercilessly short. Listening to a long baseball game, the Belmont Boulevard's owner would have been busy fumbling with batteries during the seventh-inning stretch. To make matters worse, changing the A battery was a hassle. Constant battery replacement would have been a source of displeasure to those consumers who initially paid big bucks for the radio. And there were other problems. Although the electronics had been well designed, the radio's audio performance was only fair. Use of off-the-shelf hearing aid parts (especially the ear plug) for the audio section rather than custom-designed components prevented the set from achieving its full potential.

Together, these factors paint a picture somewhat different than that of the original Belmont ad. The Boulevard's beauty was only skin-deep; it was a first-generation product that needed more work. This was not unusual, however, in the radio industry. The home radios of the early twenties were ungainly, expensive to maintain, and hard to use. But develop-

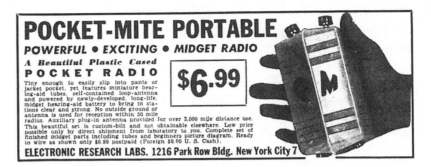

12.3 Pocket-Mite Portable radio, 1948

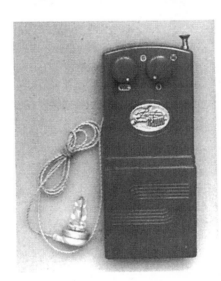

12.4 Privat-Ear Pocket Radio, 1951
(5-15/16″ high)

ment of the radio did not stop with the first models; within a few years, the technical challenges had been met by many companies. In contrast, no major radio maker—not even Belmont—devoted resources to improving the shirt-pocket portable. The entire genre was, until the coming of the transistor, relegated to the entrepreneurial fringe, which did not take it far.

In any event, I suspect that even a vastly better shirt-pocket portable would have met the same cold fate at the sales counter as did the Belmont Boulevard. Thus, explaining the failure of the shirt-pocket portable with subminiature tubes requires us to look beyond technical problems. What needs explaining, in my view, is the complete lack of faith on the part of the major companies in the marketability of shirt-pocket radios. (Recall that not even Belmont believed their set would sell.) Had any of the large companies—with their ample resources for research and development—believed that a big market existed, they might have tackled the technical problems as they had for so many other radio products. To understand this lack of faith, we must look at the place and meaning of the shirt-pocket portable in American life in the late forties and early fifties.

Today, we take it for granted that radios can be put in a shirt pocket—or even strapped on the arm—and taken anywhere. Many people regard a radio of such extreme portability as a necessity, especially if they are joggers. In the immediate post-war years, however, major manufacturers judged that a pocket radio with an earphone only (a proto-Walkman) was not apt to become a necessity for more than a few buyers.

Clearly, the tiny pocket portable had no obvious place in the lifestyle of the ordinary American adult in the late 1940s. Business people and travelers who insisted on having radio everywhere were already used to personal portables that played reasonably well, some of which could be plugged in, too. What advantage could be gained by having the ultimate in portability or the ultimate in private listening? These features would eventually be valued by adult consumers in the 1970s and 1980s, but not until there were changes in everyday activities such as exercising. Even the news or sports enthusiast who had a portable along at all times would have used it in a more public way, sharing access to information. That an adult would want to be hooked up *privately* to a radio, almost on a permanent basis, would have seemed quite odd to most people in the 1940s. And then there was the image problem.

The two most prominent features of a shirt-pocket portable, its small size and earphone, already had meanings to American consumers in the late 1940s, and those meanings were scarcely positive. Although the tiny portable radio had long been a cultural imperative among a small constituency of electronic tinkerers, extreme smallness in radios was regarded somewhat less favorably by ordinary Americans. During the Depression, Lilliputian radios had come to be associated, quite justifiably, with low price and mediocre performance. The midget plastic sets made radio affordable for almost every family, but even the poorest American had heard sets of many sizes play and knew that a midget did not sound nearly as good as a large table model, let alone a console. If ordinary midgets signified "cheapness" and inferior sound, imagine the meaning of a set small enough to slip into a shirt pocket. By size alone, then, these sets conjured up a somewhat unsavory image. During the forties and early fifties the negative meanings of a tiny radio were reinforced by consumer magazines, which were continually carping about the poor performance of all pocket radios: short life of the A battery, puny sound, and low fidelity.

Post-war consumers were decidedly not after downsized products, as several upstart car-makers also discovered. Even Crosley, the cheap radio and appliance maker, manufactured a mini-car after the war. Though it was inexpensive ($395) and got great gas mileage, it was a flop. Americans on a spending spree wanted to buy large symbols of wealth, not tiny tokens of austerity.

To the negative image of size was added, in post-war America, the separate meaning of an earphone, sometimes called an ear plug. These miniature reproducers were first widely used in hearing aids, which by the mid-forties were becoming ubiquitous. Thus, an ear plug conveyed an image of the infirmities of old age. How many middle-aged adults would have been eager to embrace an artifice that seemingly proclaimed, "My body is wearing out." (In addition, ear plugs—one size fits all—were not that comfortable and frequently slipped out of the ear.)

The hearing-aid resemblance was actually reinforced in a series of ads by battery-maker Eveready in late 1945. In *Life* and the *Saturday Evening Post,* Eveready bragged about how it had created smaller batteries for the new electronic marvels of the day—hearing aids and the (soon-to-be) pocket radio. People were pictured with these devices in their pockets, earphone plugged in. It was impossible to tell the pocket radio from the hearing aid.

As a symbol, then, the shirt-pocket-portable with ear plug was all wrong. It may have spoken of technological progress to the few people familiar with its innards, but to the ordinary middle-aged American, regarded as the "typical" consumer of household products, the shirt-pocket portable meant poor performance and raised the specter of biological decline. As a gift, a set of this sort surely would have been shunned. And to the trendsetters, a shirt-pocket portable did not make the right statement. Executives in the large radio companies, who made decisions about new products, were trendsetting consumers themselves; they could easily grasp this potential product's image problem.

But meanings are not immutable. During the early fifties, miniaturization and miniature products would begin to acquire a high-tech image as a result of the missile race and the advent of the transistor. Ever so slowly, the pocket radio would become a more desirable product as a sign of modernity. (Even so, the shirt-pocket portable would not find a mass market until the late fifties, when other factors entered the picture.)

The turning point came in 1951 or 1952, when Emerson began to envision a new pocket portable. Since the late thirties, Emerson had been a leader in radio miniaturization—portables and table models. In 1938, for example, they had built a micro-midget table model that could sit on one hand; it was called the "Little Miracle." Later, of course, Emerson made its diminutive camera-style pocket portables using miniature tubes. Now Emerson executives believed that an even smaller portable would be welcomed in the marketplace.

Emerson approached Raytheon with their plans, which called for a new set of subminiature tubes. Raytheon responded with 4 low-drain tubes (.05 watt each) suitable for a superheterodyne receiver. Emerson was unmoved by the barrage of publicity about shirt-pocket radios in electronics magazines. Shirt-pocket portables were nifty, but few Americans would buy them; every radio maker—including Emerson—knew that. Emerson's idea was to make a pocket radio that looked more like a very tiny table model, a design apparently inspired by Emerson's pre-war micro-midget.

In 1953, Emerson brought out its model 747, the smallest super-heterodyne portable since the Belmont Boulevard, and the smallest pocket radio with a built-in speaker (Fig. 12.5). The Emerson 747 was tiny and light (22 ozs.), and was easily swallowed by a jacket or coat pocket. That the radio was not intended to be placed in a shirt pocket is also indicated by its horizontal format and lack of an ear-plug jack (which precluded its use as a "sportset"). The 747's micro-table-model style would significantly influence the design of later tube and transistor portables (in the United States and Japan).

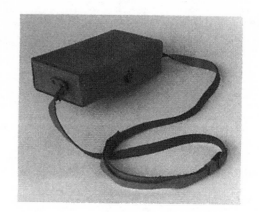

The Emerson 747 incorporated some components that evidenced significant "miniaturization" in comparison to parts used in previous pocket portables with miniature tubes. Emerson obviously had to work closely with suppliers in order to obtain adequately downsized parts. A thin plastic board substituted for the usual metal chassis, which reduced weight and furthered miniaturization, but there were no printed connections. Several Centralab integrated circuits were also used, which saved still more space. *Consumers' Research Bulletin* evaluated the Emerson 747 in December 1953. The article recited the usual litany of problems found in tiny portables:

> The Emerson 747 was sadly lacking in volume of output, and its sensitivity was barely more than sufficient to provide reception on local stations. At CR's laboratory, located about 50 miles from New York's powerful broadcasting stations, most stations were so weak that it was necessary to hold the set close to the ear to understand the announcer's words. In addition, . . . the tonal range was relatively narrow.

The article also noted that the A battery (a single C cell) lasted for only four hours—scant improvement over the Belmont Boulevard. Engaging in a bit of techno-mancy, *Consumers' Research Bulletin* suggested that the Emerson 747 "is a possible forerunner of things to come as 'miniaturizing' of circuits becomes further developed for use in consumer products." Indeed it was.

The Emerson 747 was not a virtuoso performer, but it was cute and evidently sold well at $40 (the model was continued in 1954). Dainty and purselike in its optional leather carrying-case (Fig. 12.5), the Emerson 747 was a very feminine radio. As such, it probably appealed to men as a possible gift for their wives. In addition, the Emerson 747 in no way resembled a hearing aid, and so could attract trendsetters.

The availability of the new subminiature tubes and Emerson's success with the 747 stimulated a few other manufacturers to bring out similar models in the micro-table-model style. Westinghouse, for example, offered its H508P4 in 1954. This was a poorly designed set, way too heavy for its size, especially since it was promoted in ads as a "Gift for a lovely girl." In simulated pearl finish, it sold for $39.95; in "Gray, Tan, Green, or Coral," a mere $34.95. Neither style, however, was a big seller.

The Motorola "Pixie" 45P (Fig. 12.6a) and Silvertone 4212 (Fig. 12.6b) also appeared in the mid-fifties; both contained one miniature tube as well as three subminiatures. The Pixie (of 1955 and 1956) was a modest sales success.

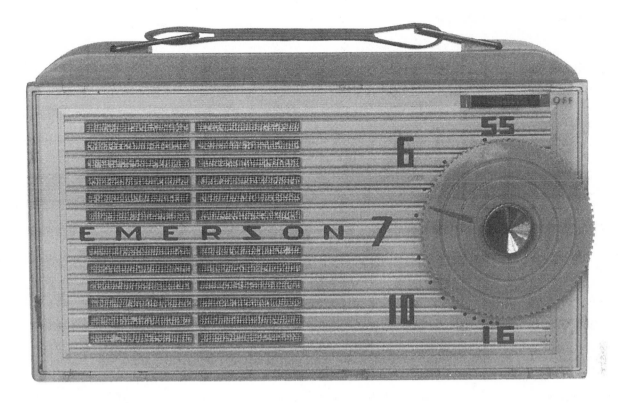

12.5 Emerson 747 and carrying case *opposite left*, 1953 (radio shown actual size). Tubes in this radio are 1v6, 1AH4, 1AJ5, and 1AG4

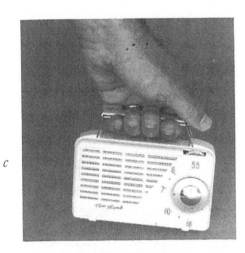

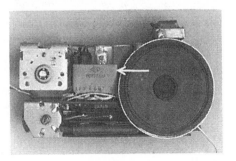

a

b

c

d

12.6 Several radios with subminiature tubes. *a*, Motorola Pixie 45P1, 1955 (6-1/2" wide); *b*, Silvertone 4212, 1954 (6-1/8" wide); *c*, Automatic Tom Thumb 528, 1955; *d*, interior of Automatic Tom Thumb 528 (arrow denotes Centralab integrated circuit)

The most important of the 747 clones was brought out by Automatic Radio of Boston in 1955. The Automatic "Tom Thumb" 528 (Fig. 12.6c) sold for $34.95 and played about as well as the Emerson 747. It was a very petite set, slightly smaller than the Emerson, and contained a new technology, the printed circuit board. In a printed circuit board the conventional components are attached to (or poke through) a plastic board on which metallic connections have been printed. The use of this postwar technology began in home radios in 1951, with Motorola the leader. By the mid-fifties, when circuit boards entered the Automatic 528, the technology was gaining a firm foothold in the industry. Its major benefit, of course, was the reduction of assembly labor (and costs); but, as the Automatic 528 showed, printed circuits also facilitated miniaturization—especially when parts were crammed close together. Transistor radios would eventually make extensive use of printed circuit boards. The micro-table-model genre of pocket radios continued to flourish in the late fifties. However, subminiature tubes had been replaced by transistors.

Lee de Forest, of audion fame, prognosticated in 1930 that the vacuum tube would never be replaced: "[It] must always be the foundation [of broadcasting], for I cannot conceive of any other method of detecting and amplifying delicate radio and audio energy than the simple vacuum tube." Eighteen years later Bell Laboratories, the prestigious research arm of AT&T, revealed at a news conference in the spring of 1948 one of the best kept secrets of the post-war period. A team of Bell physicists had invented a new device, the transistor, that could also amplify. Lee de Forest had been wrong.

Dwarfed by an eight-foot-high scale model of their creation, Bell Laboratory functionaries proudly predicted that the transistor would soon replace the vacuum tube in many applications. This point was driven home dramatically by several demonstrations, including the playing of a portable radio whose entire complement of tubes had been replaced by transistors. According to Bell Labs' own account of this show, "Reception was excellent." Once again, a portable radio had sent a signal of technological progress. AT&T's transistor show-and-tell demonstrated to engineers and managers in radio companies that the tube portable was a technological dinosaur, teetering on the brink of extinction.

The popular press initially paid little attention to the transistor, relegating coverage to small articles tucked away on back pages. *Electronics,* however, made the transistor its cover story in September of 1948. Though physicists and engineers were enthusiastic about the transistor's potential, no one then foresaw that, at the heart of computers, transistors would decades later cause an information revolution nearly as far-reaching as that wrought by radio itself.

The development of the transistor at Bell Labs was no accident. Rather, it was the purest of pure science harnessed by one of the world's largest corporations having relentlessly practical concerns. As early as the middle 1930s, Mervin Kelly, director of research at Bell Labs, anticipated that the mechanical relays in all telephone exchanges would one day limit the telephone system's growth. A vision eventually emerged of electronic switches and amplifiers that did the work of tubes but without their bulk and power consumption. In these hypothetical devices there would be no filament, and current would flow in a solid material instead of in a vacuum.

Mervin Kelly was not the first person to conceive of the idea of a solid-state amplifier. With the advance of solid-state physics during the Depression, such an amplifier had become a cultural imperative. Many people—in the United States and abroad—had begun to work on the problem. In fact, several promising designs had been patented in Germany in the 1930s, though no functioning devices had been made. Among those with the idea, then, Kelly was ideally situated to begin exploring its feasibility experimentally. With unmatched facilities and millions of dollars at his disposal, he could mount a credible research and development program.

Kelly assembled a team of scientists in 1939 to tackle the solid-state amplifier. These early experiments, interrupted by the Second World War, were unsuccessful, in part because materials of sufficient purity could not be obtained. During the war, radar-related research led to techniques for producing ultra-pure germanium and silicon for crystal detectors; physicists believed that these materials could also be used in solid-state ("crystal") amplifiers. Immediately after the war, effort was renewed to realize Kelly's vision. Bell Labs put together a team of physicists headed by William Shockley, a theoretician who was confident that a solid-state equivalent of the triode was possible in principle. The other team members were John Bardeen, also a theorist, and Walter H. Brattain, a gifted experimenter.

The exciting interaction between theory and experiment led directly to the transistor. On December 23, 1947, the group successfully demonstrated the point-contact transistor, which consisted of two closely spaced wires on a germanium crystal. A solid-state amplifier, about the size of a pencil eraser, had been created, which was christened the transistor. For their stunning feat, Shockley, Bardeen, and Brattain were awarded the 1956 Nobel Prize for Physics. (Shockley and Bell Labs in the early fifties also invented other kinds of transistors.)

Though a remarkable scientific achievement, invention of the transistor was but the first of many steps needed to put into phone equipment the practical device needed by AT&T. Mass-production of the transistor presented devilish problems. Transistors often did not work or, if they did, their characteristics differed from those desired. Assembly of transistors turned out to be a hit-or-miss process often described as an art or "black magic." At the end of the assembly line (which was not automated) a worker tested each new transistor; more often than not, the bucket of "rejects" had four or five times as many transistors as the "good" pile.

These production problems were solved slowly. Not until late 1951 did Western Electric, manufacturing arm of AT&T, begin serious production of the transistor. Raytheon actually began producing small numbers of transistors commercially in late 1949, but true mass-production did not begin at this company until late 1952. By that time, a few other firms were also selling some transistors (at about $18 each).

After achieving its breakthrough in mass-production, Raytheon began pushing hearing-aid companies to convert to transistors. Transistor hearing aids could be operated at less than one-tenth the cost of the tube models. Though reluctant to give up the very successful subminiature tubes for an unproven substitute, hearing aid companies did accede to Raytheon's pressure. And so the first commercial product to use the transistor was the hearing aid, examples of which reached the market before the end of 1952.

Needless to say, experimenters were itching to get their hands on the new (but very expensive) crystal amplifiers. In January 1950, Rufus P. Turner, an amateur, published an article in *Radio and Television News,* which noted that "the recent appearance of the transistor on the commercial market opens new fields to the radio experimenter." Indeed it did. In the article, Turner presented plans for a three-transistor radio that employed Raytheon's CK703 transistor. Rufus Turner was busy in 1950. In June, he published another article, this time in *Radio-Electronics,* describing several simple circuits that also used the CK703, including a new radio. Other home-brew radio circuits soon followed, especially after the debut, in early 1953, of Raytheon's CK722 transistor, the first commercial transistor available—at about $8 each—to experimenters in large numbers.

Because commercial shirt-pocket portables were on the market already, the experimenter pushing the frontiers of transistor applications sought more challenging projects. Another longstanding cultural imperative had been the wristwatch radio. As Gernsback noted in 1951, the transistor would make possible "during the next few years an excellent superheterodyne radio receiver, no larger than a wrist-watch."

Experimenters were soon at work on the wristwatch radio. An early example was made in 1952 by two Western Electric engineers in their spare time. It measured 1-1/2" × 2" × 3/4", and employed four transistors. Although the radio contained some miniature components, it did not include the speaker, battery, and antenna. Why had the engineers made the radio? It was a gift for Chester Gould, creator of Dick Tracy. The radio was also intended to show, the article in *Radio-Electronics* stressed, "what transistors may be able to do at a not-too-distant day."

Another early wrist radio, described in *Radio and Television News* (July 1953), was made by M. E. Quisenberry. A one-transistor radio with an earphone, it was billed as "an interesting toy for the youngsters that will fascinate adults as well—a wristwatch 'Dick Tracy'–type radio set." Though pretty small, with batteries on the outside, it was not a very elegant piece of wristwear. Closer to the ideal was a wrist radio put together in 1953 by the U.S. Signal Corps Engineering Laboratories at Fort Monmouth, New Jersey (Fig. 12.7). This was a three-transistor set that used a printed circuit board and a very tiny mercury battery. The antenna was a short wire that ran up the sleeve "comic-strip style."

The Signal Corps and other U.S. military agencies were the largest market for transistors in the fifties; more than $50 million was poured into direct subsidies of transistor-making companies. The products were used in countless experimental projects, like the wrist radio, and underwent extensive testing under varied conditions. Ordinarily, military ap-

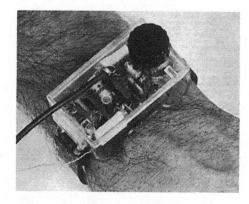

12.7   U.S. Army Signal Corps wrist radio, 1953

plications of transistors had a low public profile, but every now and again a project was publicized in order to call attention to the exciting, cutting-edge work being done in the military labs. At this time, of course, the wrist radio was the perfect product for such a technological display; the tiny set was written up in *Science Newsletter* and *Radio and Television News,* which doubtless made readers more kindly disposed toward the Signal Corps.

It would be some years before the transistor replaced tubes in most electronic products, which Bell Labs had so brashly predicted in 1948. The reason was simple: transistors were still very expensive, with an average price in late 1953 of around $8 each (an ordinary radio tube was less than $1). But each year the price fell, and by 1960, the average price descended well below a dollar. In anticipation of declining transistor prices, RCA in late 1952 put on a show-and-tell to the trade, displaying a variety of experimental transistorized products. Foremost among these was a five-inch, battery-operated portable television having thirty-seven transistors, an eleven-transistor car radio, a transistor ukelele, and, of course, a portable radio. Soon thereafter, a circuit diagram for a transistor portable was distributed by RCA to its patent licensees (most radio and TV companies).

Much of the original development of transistor production techniques took place at the established electronic companies, which also made tubes, such as RCA, Philco, and Raytheon. However, many new companies moved in quickly, and these upstarts became significant players in the solid-state arena. Shockley himself founded the first solid-state firm in the area of northern California near Stanford University, his alma mater, which is now known as Silicon Valley. Texas Instruments was another company new to component manufacture, and, as Michael Wolff recounts in "The secret six-month project," they had big plans for producing and commercializing the transistor.

Texas Instruments obtained a license from Bell Labs, and by the end of 1953 was mass producing the tiny solid-state amplifier. The following spring, Patrick E. Haggerty, executive vice-president of Texas Instruments, assigned a team of about a dozen engineers to design a product for consumers. Texas Instruments, however, had no expertise in consumer electronic products. Unfettered by either the wisdom or prejudices of established radio companies, they decided to make a shirt-pocket portable. Their goal was to mass-produce this radio in time to reach the Christmas market. Within a month a six-transistor prototype had been created, which fit into the empty case of an Emerson 747.

The next task was to find a radio company that would manufacture their set by Christmas, *using Texas Instruments transistors.* This turned out to be more difficult. In the words of S. T. Harris, Texas Instruments' marketing director, "I personally contacted every major radio manufacturer in the United States by phone, telegram, or letter and got no encouragement from any of them." The Texans received a chilly reception at RCA, General Electric, Philco, and so on because these companies were already working on transistor radios of their own design (and many of them were making transistors as well). Texas Instruments was just one of many companies that, following RCA's lead, had appreciated the possibilities of a transistor portable radio. In addition, the established radio companies were not interested in a shirt-pocket design; these, after all, were not serious radios, and Americans would not buy them.

Texas Instruments finally made contact with a small but reputable electronics company in Indianapolis, I.D.E.A., and the latter was eager to take on the challenge. The president of I.D.E.A., Ed Tudor, was confident that the shirt-pocket radio's time had come. After all, this was 1954, when the Cold War was rather frigid; the Soviet Union now also had *the bomb,* and Americans everywhere were jittery. School children practiced "drop drills," and all radio dials got little CD (Civil Defense) triangles at 640 and 1240 kHz, which indicated the only stations that would be broadcasting in the event of a "nuclear attack." Portable radios were essential in such emergencies; the smaller the better, as Hugo Gernsback had already noted in 1950, when he called on manufacturers to produce one-pound "emergency receivers" for "the atomic age." Ed Tudor had precisely this market in mind when he eagerly joined forces with Texas Instruments. His sales goal was "20,000,000 sets in three years."

I.D.E.A. engineers streamlined the circuit somewhat, reducing the number of costly transistors to four. New, miniature components had to be designed and their manufacture farmed out to other companies. An industrial design firm from Chicago (Painter, Teague, and Petertil) created the radio's distinctive case, which was reminiscent of Paul Beshore's pocket crystal set. Under great pressure from the approaching deadline of mass production by November, the engineers and designers tinkered with the case and chassis configuration, finally getting everything to fit. On October 18, 1954, production of the new radio—in time for Christmas—was announced to the public. Called the Regency TR-1, it was the world's first shirt-pocket portable radio—*with transistors* (Fig. 12.8, right). Predictably, some Regency ads boasted that the TR-1 was the world's first pocket radio (Fig. 12.8, left).

The Regency TR-1 fit into shirt pockets comfortably, its plastic case being only 5″ high, 3″ wide and 1-1/4″ thick. Containing four transistors and a printed circuit board, the radio weighed—including battery—only eleven ounces, half as much as the Emerson 747. Apparently aware of their creation's fragile plastic shell, the makers of the TR-1 offered a leather carrying case as an accessory; an earphone could also be purchased. The Regency TR-1 may have fit all pockets but, at $49.95 (plus accessories), it did not fit all pocketbooks.

What soon came to be the transistor's greatest advantage over tubes—vastly lower power consumption—was already evident in the TR-1. With a 22.5-volt battery, costing $1.15 in 1955, the TR-1 would play for between twenty and thirty hours, which was much better than the tiniest tube portables (three to five hours) before a change of the A battery was needed. Better still, overall costs of operation were lowered, from seven to twenty-one cents per hour to about a nickel. Battery economy of the TR-1 was a palpable improvement over the smallest tube sets, but within months the transistor radio would break with ease the penny-an-hour barrier.

Despite its low operating costs, the Regency TR-1 was not received

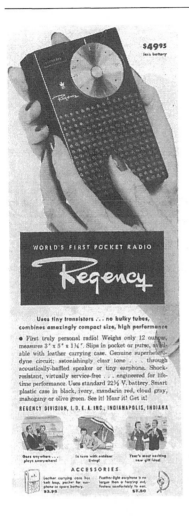

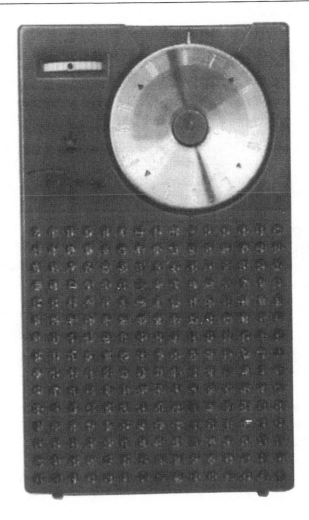

12.8   Regency TR-1: The world's first transistor radio, 1954 (*right,* radio shown actual size)

with enthusiasm by consumer magazines. In its review of July 1955 *Consumer Research Bulletin* remarked on the TR-1's poor tone quality, unpleasant distortion at high volume, low power output, and excessive background noise. *Consumer Reports* also told a tale of woe:

> Though its transmission of speech was adequate under good conditions, its music transmission was quite unsatisfactory under any conditions. Little of the bass notes came through. The signal hissed even on strong stations and tended to whistle and squeal at several spots on the dial. At low volume the sound was thin, tinny, and high-pitched and at higher volume the distortion increased.

Those seeking transistor radios were counseled by *Consumer Reports* "to await further developments before buying." Clearly, the Regency's small size and low battery drain were achieved at a price.

Although not a stellar performer (the Emerson 747, despite having lower volume, sounded *much* better), the high-tech Regency TR-1 attracted the trendsetters and electronics buffs. Within a year, sales approached 100,000 sets. This was a respectable figure for a novelty item, but the mass market was untouched; the multi-million demand for emergency receivers never materialized. It is doubtful that Texas Instruments had any earnings from the pocket radio project; their crash development program had cost more than a million dollars. I.D.E.A. made a number of additional Regency models, but in several years—under financial duress—quit the pocket radio business. In *Electronic Inventions and Discoveries,* Dummer went so far as to conclude that the TR-1 was "not a commercial success." Certainly its sales did not meet expectations.

Nonetheless, Texas Instruments did earn a huge dividend from the Regency TR-1, for this shirt-pocket portable demonstrated to all in the know that the transistor age had arrived. Haggerty somewhat later acknowledged that the real purpose of the first transistor radio was technological display: "It seemed clear that a dramatic accomplishment by Texas Instruments in the field of semiconductors was needed to awaken potential users to the fact that the devices [transistors] were usable now, not some years in the future, and that we were ready, willing, and able to supply these usable devices."

Originally, the Regency TR-1 was intended to showcase silicon transistors, which Texas Instruments succeeded in making in 1954, ahead of the competition. Silicon transistors had a number of advantages over the germanium ones then in use, and firms across the country were working on them feverishly. It would have been a magnificent coup, but the silicon transistors just were not ready in time for the TR-1. Even so, the first transistor radio did have the desired effect.

The importance of the TR-1 in stimulating demand for transistors was especially clear at IBM. Thomas J. Watson, Jr., head of IBM, bought several hundred of the new sets. Not only did he give TR-1s to various IBM executives, but he used them to goad his engineers into building computers with transistors. Comfortable with the familiar tube circuits, computer people had been reluctant to embrace the new technology. An impatient Watson finally ordered that IBM build no more machines with tubes after June 1, 1958. Whenever engineers complained about the decision to Watson, he just gave them a TR-1. The message was clear. Incidentally, Texas Instruments supplied transistors for those new IBM computers and by 1960 had sales of $200 million—almost ten times its 1953 performance.

Although the Regency TR-1 led to the more rapid application of transistors to computers and to many other electronic devices, and helped make Texas Instruments a giant in the semiconductor field, in 1954 and 1955 few beyond the trendsetters and electronics buffs had much interest in a petite portable that could not reproduce music. That situation would change in a few years, when transistor shirt-pocket portables sounding little better than the TR-1 would sell annually by the millions. To understand the eventual resounding success of the shirt-pocket genre we must look at changes in popular music that took place in America of the fifties. The tiny pocket portable would not enter homes—and pockets—in large numbers until America underwent a musical revolution.

13

"She's My Transistor Sister"

WHEN THE REGENCY TR-1 DEBUTED IN LATE 1954, Elvis Aron Presley still made a living driving a truck. Rock and roll was not yet a household word, much less a big business. Largely because of Elvis, that would soon change.

Elvis was an undeniably talented singer and entertainer, but so were many of his contemporaries. What distinguished the man from Tennessee was his unique blend of black and country music traditions. For the first time a nationwide audience—mostly white, novelty-seeking youth—was exposed to black music's intensity and rhythm. Responding to the beat, white kids and brown kids and black kids wanted to dance and shake their hips like Elvis, for this music touches emotions that are colorblind. Though many apprehensive adults believed that rock music and dancing were expressing sexuality, rock's origin was, ironically, in the church, in the religious music of poor southerners. Elvis himself at first wanted to be a gospel singer, but failed the audition.

Needless to say, the possibility that rock music aroused sexual passions was not lost on self-appointed guardians of mid-fifties morality. Popular singer Frank Sinatra, whose crooning had enchanted forties teens, scorned the new music. The *New York Times Magazine* reported his harsh assessment: "Rock 'n roll smells phony and false. It is simply played and written for the most part by cretinous goons and by means of its almost imbecilic reiteration, and sly, lewd, in plain fact, dirty lyrics . . . it manages to be the martial music of every side-burned delinquent on the face of the earth."

Though Sinatra lacked insight into the musical changes that seemingly threatened his own livelihood, at least his condemnation of rock was not overtly racist. Others were less restrained. One Asa E. Carter, leader of the North Alabama Citizens Council, charged in 1956 that rock and roll was an effort to force "Negro culture" on southern whites. This music, he said, "is the basic, heavy-beat music of Negroes. It appeals to the base in man, brings out animalism and vulgarity." His group was formed to persuade jukebox owners to eschew rock records. Doubtless, the friendly persuasion of Asa and his ilk sometimes lacked subtlety, and such intimidation was effective. In San Antonio, for example, rock and roll was banned from swimming-pool jukeboxes in 1956 because "its primitive beat attracted 'undesirable elements' given to practicing their spastic gyrations in abbreviated bathing suits."

Because rock did not follow the conventions of traditional popular music, such as lush orchestration and lyrical refinement (and had obvious sexual overtones), it was strange and profoundly disturbing to many adults. Adding to the anxiety were movies of the mid-fifties—*Blackboard Jungle, Rebel Without a Cause, The Wild One*—that stereotyped teenagers as rock-crazed juvenile delinquents.

However, opposition to Elvis and rock faded rapidly in the entertainment business as people began to appreciate its possibilities for commercial exploitation. During the mid-fifties, both Steve Allen and Ed Sullivan hosted competing television variety shows on Sunday night. These were "wholesome" programs that appealed to the entire family. Occasionally, a pop singer would even be sandwiched between the comedians, tap dancers, and jugglers. Sullivan, however, was no Elvis fan, proclaiming early on that the Tennessee sensation would never appear on his show. And so Steve Allen was the first to sign Elvis. He did one performance, in July of 1956, and the effects on ratings were impressive. Not long afterward, Sullivan swallowed his scruples and paid Elvis $50,000 for three appearances. It was money well spent. The first show captured 82 percent of TV viewers, though Elvis was shown only from the waist up. RCA signed Elvis in late 1955 and by the next year sales of his singles and albums were contributing significantly to this giant corporation's profit picture. Eventually, even Frank Sinatra sang a different tune and included Elvis in one of his television specials. Soon teenagers could watch their peers dancing every afternoon to the new music on Dick Clark's *American Bandstand.*

Elvis' swiveling hips obviously held an attraction for girls, the gender that did most of the rock record buying (some estimate 90 percent of the forty-fives sold), but rock rhythm and lyrics also appealed to boys. Between the time of puberty and the time of marriage, American youth—not as likely to be working in factories or on farms—had a great many years on their hands during which to explore sexuality and love. Rock music, written mostly by males, articulated these themes, sometimes poignantly. *Lonely Boy,* sung in 1959 by Paul Anka, expressed these feelings:

I'm just a lonely boy, lonely and blue
I'm all alone, with nothing to do
I've got everything, you can think of
But all I want, is someone to love.

Rock music understood the trials of youth in ways that adults did not, and so appealed to both sexes.

The father of rock and roll was not Elvis but Bill Haley, whose "Shake, Rattle and Roll" reached the top ten in 1954. In July of the following year "(We're Gonna) Rock around the Clock," by Bill Haley and the Comets, theme song from *Blackboard Jungle,* topped the Billboard chart where, as the first rock record to hit number one, it stayed for eight weeks. According to many music historians, this event marks the birth of the rock era. Regrettably, Bill Haley and the Comets had no other great hits.

But Elvis had staying power. His first RCA single, "Heartbreak Hotel," went to the top of the charts in April 1956; within the next year, Elvis had five more number ones. By the beginning of 1958, other rock singers had also reached the top of the chart, including The Platters, Paul Anka, The Crickets, Jimmy Rodgers, the Everly Brothers, and Sam Cooke. Rock was on a roll. Forty of the top sixty singles in 1957 were rock and roll; the best seller was "All Shook Up," an Elvis tune that sold over two million records.

Station managers soon figured out that teenagers themselves comprised a large radio audience that could be targeted with rock and roll programs. The first major radio station to air rock was WINS in New York in 1955. "Disk jockey" Alan Freed played the platters during a special time slot and gave the new music its name, "rock-and-roll." Other stations quickly followed suit. WHB in Kansas City became the first to play rock exclusively. Within a few years a slew of "top 40" stations across the country was serving the youth market. In the privacy of their rooms, kids could listen to the all-rock stations on their own table radios. In more public areas, however, parents and children became locked in mortal conflicts over which stations the family radios would play, with parents usually winning.

Small transistor portables, equipped with ear plugs, elegantly solved the rock and roll problem. By bestowing these radios as gifts, parents could wall off the offending music, insulating themselves from its erotic drives. Youth, of course, was unencumbered by the negative meanings that tiny radios once had for their parents and grandparents. Teenagers soon discovered that transistor portables, especially the shirt-pocket variety, gave them and their music unprecedented mobility. Eventually, teenagers came to believe that they were screening off the rest of the world and creating their own. The shirt-pocket portable or, simply, *the transistor* (as it was called then) became a metaphor for freedom and independence; the right to express, in music and in things, the style and tastes of youth. The tiny transistor radio had become the symbol of a generation. Somewhat later, R. J. Gleason, in *American Scholar*, remarked that "This was, truly, a new generation—the first in America raised with music constantly in its ear, weaned on a transistor radio." Rock record producers even shaped their music so as to sound a little less dreadful through the miniature speakers. By the time rock and roll was here to stay, in the waning years of the decade, the Japanese had put inexpensive portables within the grasp of most teenagers (see Chapter 14).

Radio makers had no inkling that teenagers would become the ultimate consumer of small transistor portables. After all, the first ones were expensive; thus, ads in magazines like *Holiday* and *Colliers* were aimed at an upscale adult market (Fig. 13.1). But many businessmen and sportsmen and so on had acquisitive teenage children, eager to try out dad's new toy. Possibilities emerged for a musical accommodation between the generations.

Radio companies eventually figured out who ended up with their small transistor portables. Zenith, for example, advertised in *Senior Scholastic;*

and, in the late fifties, transistor radio ads had joined those for girdles, lipstick, and portable phonographs in *Seventeen*. Transistor radios also turned up as a desirable product in *Boy's Life;* an ambitious youth could even earn a radio by selling greeting cards.

Though in the entertainment industry rock had become established, detractors persisted. *Newsweek* even wrote an obituary for the new music in 1959, claiming that "raucous rock 'n' roll seems to be on its way out, after hogging the air waves for four years." The article described how radio stations had begun to realize "that their adult audiences are fed up with the bleating cacophony of rock 'n' roll." Assuming that teenagers purchased little beyond records and soft drinks, station personnel were writing rock out of their play lists. Interviews with program directors and disc jockeys from stations in six cities seemingly supported *Newsweek*'s generalizations. Typical was the comment from the program director of KRKD in Los Angeles: "We're back to the old-fashioned idea of people tuning in to listen to good music. The net result is more and better business."

Rock's obituary, like Mark Twain's, was somewhat premature. *Newsweek* had failed to appreciate that these changes in programming were part of a larger package of changes that affected radio in the 1950s, not just in response to rock but to television.

The adults and children who, in the thirties and forties, gathered around the family radio to hear "Amos 'n' Andy" by the early fifties were watching them on a twenty-one-inch screen. The centerpiece of the living room was no longer the console radio but the console television. The "one-eyed monster," the "boob tube," the "idiot box," as later critics would call TVs, had replaced radio as the focus of family entertainment, especially during the evening (Fig. 13.2).

As in radio, the major networks dominated television programming. Consequently, radio and TV networks were competing for the same prime-time audience, but television was the decisive winner. In the decade from 1946 to 1956, the maximum family audience for network radio dropped 90 percent, from 25 million to 2.56 million, while income fell 50 percent. Surprisingly, independent radio stations were doing just fine. In fact, in the decade following World War II, the number of U.S. stations almost tripled. What was their secret?

13.1   Regency ad targeted at upscale consumers in *Holiday* magazine

13.2   By the early fifties, TV had taken over the living room, focus of family entertainment in the evening (Magnavox Chippendale TV, 1952)

Since the 1920s, many unaffiliated stations had discovered a format that appealed to a sizable segment of the radio audience. This format assumed that radio's most important entertainment function during daytime was to provide background music. *Time* magazine in 1955 identified the "new average radio listener": "He treats his radio like a constant companion who is pleasant to have around but can be comfortably ignored. The dial twister listens intermittently while getting up, before going to sleep, while shaving, eating, working around the house, or driving."

Independent stations thrived with this format in the 1950s because, for an ever-larger segment of the audience, radio had become background sound to other activities at home, at work, and at play. When Americans sought more engaging entertainment, especially in the evening, they turned to television.

Writing in *Atlantic Monthly* in late 1959, long-time broadcaster, William O'Hallaren, lambasted the "formula radio" that was out-competing the networks. His description of the successful but impoverished radio format—"top music and news"—still seems current today:

> The "top music" of the gypsy formula means the thirty or so numbers leading one of the weekly surveys of taste in this field, and a considerable number of stations actually refuse to use any other music. The "good news" of the formula is hardly more than a pious afterthought and generally consists of a five-minute newscast each hour . . . bland headlines ripped from the radio wire of the Associated Press and delivered in a confident shout.

Network radio finally began to adapt to the altered realities of the television age by reducing its hours of programming, giving more time to local control. Eventually, its most distinguished voices, including Edward R. Murrow, Walter Cronkite, and Eric Severeid, moved to television, as did its most successful programs. In this atrophied form, network radio still survives.

The most notable change in radio programming of the late fifties was based on the tastes of generations. Some stations, as *Newsweek* had noted, expunged rock from their play lists, wooing an older and ostensibly wealthier audience. Other stations adopted rock programming exclusively and, as a result, thrived. In a more affluent America, teenagers had money to spend—$10 billion a year by 1960—and they also influenced their parents' purchases. By the time KRKD in Los Angeles dropped rock, it mattered not; teenagers had already tuned out, listening instead to all-rock stations KFWB and KRLA. In a remarkable journalistic lapse, *Newsweek* had entirely missed the growth of stations playing only rock. In any event, with the coming of rock-only radio during 1955 to 1960, the musical isolation of the generations was complete.

A generation of Americans could now use their own (pocket) radios to listen to their own music. More than any other factor, the establishment of all-rock stations created the mass market that had long eluded the shirt-pocket portable.

One of the transistor's inventors reportedly quipped in later years that if he had known that their device would give millions of teenagers the means to play rock and roll, he might have rethought the entire proposition. Though Bell Labs' forecasts for the transistor had mixed success, at least initially, forecasting the future remained a respectable sideline of scientists and engineers during the 1950s. It was important for the techno-mancers to speak directly to the people and to policy-makers because, in the aftermath of Hiroshima, the terrible potential of science and technology to destroy was apparent to all. Moreover, the writers of science fiction no longer served uniformly as propagandists for a high-tech future. Many films of the fifties, for example, explored technology's dark side. Technology could create dystopias as well as utopias.

Yet, around them in everyday life, Americans of the fifties saw ample evidence of technology's benefits. Indeed, many cultural imperatives of earlier times had come into being, from Bellamy's music box to Henson's aerial steam carriage. Homes fairly bulged with refrigerators, dishwashers, clothes washers and dryers, televisions, and countless other products of modern technology. Even the portable radio was commonplace. Technology was beginning to catch up with techno-mancy. It was obvious to ordinary Americans that technology itself was not inherently harmful, though it could be misused. In an enlightened society, most believed, technology's negatives would be held in check. Among the masses, enthusiasm for technology did not wane; what is more, unrealized cultural imperatives, such as space travel, were sold to larger constituencies and served as stimuli for further technological development.

The century's greatest electronic techno-mancer, Hugo Gernsback, in the fifties was still busy prognosticating miniature things appropriate for the transistor age. In 1952, before any computers were transistorized, the indefatigable Hugo had foreseen desk-top computers. A few years later he envisioned atomic-powered electronics with highly integrated components, shirt-pocket and wristwatch televisions, and "television eyeglasses." In a 1955 editorial in *Radio-Electronics* he contemplated the phone of the future: "The day will come when all of us will be able to carry a miniature compact dial phone in our pocket for instant communication with the entire world."

Gernsback wrote his last science fiction novel in 1958–59 (though it was not published until 1971). *Ultimate World* told the story of two advanced civilizations that realized their capacity for mutual annihilation. Despite its transparent commentary on the missile race, the book was not all bleak: Gernsback described the endless bliss of sexual intercourse in a weightless condition. In 1958 the Hugo Awards were established in his honor, and remain to this day the highest recognition that an author of science fiction can achieve.

Sarnoff, still in command at RCA (he did not retire until 1969), was for the first time playing a catch-up game, trying to master computers. He was aware—before many others—that computers could revolutionize the world of white-collar work. As always, he wanted RCA to be the industry leader in the new technology. But RCA never caught IBM and eventually bowed out of the computer business after suffering immense losses.

Looking ahead, Sarnoff's vision was increasingly clouded by Cold War concerns. In one oft-quoted statement, he remarked that "the best way to prevent a hot war is to win the cold war." In correspondence with Vice President Richard Nixon, Sarnoff—respected elder statesman of the electronics industry—set forth strategies for winning the Cold War. Not surprisingly, these mostly involved new weapons systems heavily dependent on electronics. Indeed, selling high-tech, high-priced gadgetry to the military was good business for RCA. But RCA was not alone; profits of many electronics companies were pumped up to record levels in the late fifties by military spending.

The modern symbiosis between the electronics industry and the military emerged during World War II. Whereas radios on a pre-war destroyer had 60 tubes, electronic devices of all kinds in the destroyer of 1952 had upped the tube count to 3,200. This pattern repeated in other ships, planes, tanks, and in countless new applications, as electronics infiltrated every nook and cranny of military operations.

During the Korean War, the U.S. government encouraged companies to expand capacity rather than convert from civilian to military goods. When this "guns and butter" war ended in 1954, military contractors—including many electronics firms—were stuck with excess capacity at the same time that a severe recession loomed. But a solution was soon found.

In 1955, a year after the war's end and the advent of the frozen TV dinner, the Russians got the H bomb. Senator Joe McCarthy's witch hunts had already stirred much of the nation into an anti-communist frenzy, and so many people considered the "godless Soviet communists" a danger to capitalist America. The Cold War, then, was a way to keep factories humming and workers working while ostensibly parrying grave threats from a nation, a former ally, that was unable to feed its people. With its Pentagon booster aboard, the U.S. consumer society—and the electronics industry—lunged ahead on a new trajectory.

Many electronics firms, taking a bath on color TV and impatient with fickle consumers, eagerly expanded their military divisions in the late fifties. Military procurement seemed to offer a stability lacking in the home electronics business. These moves would lay a foundation in many companies for jettisoning consumer electronics altogether in later years.

The Cold War—the missile and space races, in particular—pushed hard on the frontiers of electronic technology. In contrast to Russia's rocket engineers, who built powerful but not so sophisticated missiles, America's rocket engineers favored missiles with less muscle that were electronics-intensive. Because U.S. missiles could loft only limited payloads, miniaturization was promoted across a broad front. By mid-1958, for example, U.S. satellites contained solid-state circuitry. Not only did the U.S. military continue to be the largest purchaser of transistors in the late fifties, but it also funded industry programs to develop new technologies of micro-miniaturization and circuit integration.

In theory, solid-state devices were capable of almost infinite miniaturization and integration, and in practice great successes came quickly. The most significant development, created by both Fairchild and Texas Instruments at the end of the decade, was the modern integrated circuit, built (eventually) on a silicon chip. Such components contained transistors, resistors, and capacitors, and were the building blocks of computers—mainstay of missile guidance systems and satellites. Clearly, the United States was the undisputed leader in electronic miniaturization, and would remain so for decades.

These new technologies, however, did not pass readily into the consumer sector. Not only were silicon integrated circuits expensive, but few executives forecast much demand for radios the size of a grape or TVs the size of an apple. Well into the late sixties, then, the major market for silicon integrated circuits remained the U.S. military. (The first consumer product to use one—in 1964—was a Zenith hearing aid.) Although integrated circuits were on the cutting edge of miniaturization, discrete components for transistor products were not ignored. During the late fifties a host of firms incessantly advertised in the pages of *Electronics* and other trade magazines their "subminiature" resistors, capacitors, transformers, and so forth. Though used mainly for military, industrial, and commercial applications, these expensive parts were also suitable for radios and TVs. Presumably, prices of the mini-components would have dropped had there been much demand for them in consumer products—but there was none. In the late fifties, radios and TVs simply did not require parts miniaturized beyond those present in the Regency TR-1.

Although the Regency TR-1 was the world's first consumer transistor radio, it was not first by much. Number two, the Raytheon 8-TP-1 (Fig. 13.3, left), came out early in 1955. Radio makers had kept a close eye on transistor prices and were gearing up to make portables. In 1955, as transistors descended to $3 each, most companies put their plans into action. From 1955 through 1959 there was a veritable explosion of U.S.–made transistor sets (Fig. 13.4a–w).

Most companies at first eschewed shirt-pocket portables, believing that the public would not accept their impoverished sound. Raytheon, for example, was not interested in putting out the first or the smallest transistor radio, merely one that played well. Perhaps recalling the Belmont Boulevard's debacle, Raytheon engineers designed a small lunch box set; the 8-TP-1 had eight transistors and a four-inch speaker—the same size as most table models. It cost a princely $85 and was an impressive performer. Like the Raytheon radio, many early transistor portables were made in the lunch-box style, with leather, plastic, or metal cases. Most of the small plastic radios were patterned after the Emerson 747, in the micro-table-model style; a few approached shirt-pocket size (Fig. 13.5). However, U.S.–made shirt-pocket portables were uncommon until after the Japanese, in the late fifties, showed the existence of a huge market of teenagers.

Doubtless inspired by the widely publicized Signal Corps set, LEL, a tiny firm in Copiague, New York, in 1956 manufactured the first commercial wrist radio. The three-transistor set was powered by five button-size mercury cells. Though neither compact nor elegant, this ear-plug-only radio could pick up strong local stations without an external antenna. The LEL, an obvious novelty, attracted few imitators or a consumer following, and quickly disappeared.

13.3 Two transistor radios from 1955. *Left,* Raytheon 8-TP-1, the world's second transistor radio (9-1/4″ wide); *right,* Zenith's first transistor radio, the Royal 500, 1955 (5-3/4″ high)

13.4 A sampling of transistor portables, 1956–1959, made in the U.S. with U.S. parts. *a*, Emerson 855, 1956 (7-1/4" high); *b*, Magnavox AM-2, 1956 (5-5/8" wide); *c*, Philco T7-126, 1956 (7" wide); *d*, Emerson 888 Explorer, 1958 (6-1/2" high); *e*, Emerson 888 Pioneer, 1958 (6-1/2" high); *f*, Firestone 4C43, 1958 (6-1/4" wide); *g*, RCA T1EH, 1959 (7" high); *h*, interior and exterior of General Electric P750A, 1958 (6" high); *i*, Zenith Royal 760, 1959 (5-1/8" high); *j*, Arvin 9595, 1959 (7-1/8" wide); *k*, Admiral 692, 1959 (5-1/2" wide); *l*, J.C. Penney RP-1-124, 1958 (5-7/8" wide); *m*, Motorola X12A-1, 1959 (5-7/8" wide); *n*, Sylvania 3204TU, 1958 (6-1/2" high); *o*, General Electric P725B, 1958 (8-1/4" wide); *p*, RCA 1T55, 1959 (9-3/8" wide); *q*, General Electric P808A, 1959 (7-1/4" wide); *r*, RCA 1T2F, 1959 (7-1/4" high); *s*, Philco T700, 1959 (9-1/2" wide); *t*, General Electric P746A, 1958 (6-1/2" wide); *u*, Motorola 8X26S, 1959 (7-1/8" high); *v*, General Electric P715, 1956 (6-3/4" high); *w*, Trav-Ler TR-280B, 1958 (5-9/16" high)

With the coming of the transistor, portables enjoyed a sales surge. In 1955 almost 2.5 million sets made their way into the hands (and sometimes pockets) of American consumers. Almost one-third of the portables sold in 1956 were transistor sets. By mid-1957, all viable radio companies were making them. And, in that year, transistor portables outsold those with tubes. One of the most successful early models was Zenith's first transistor radio, the Royal 500 (Fig. 13.3, right), which was manufactured from 1955 to 1959. Containing seven transistors, a crystal diode, and a speaker slightly larger than that of the Regency TR-1, the stylish Royal 500—with its nylon case—performed about as well as some pocket portables with tubes. It was rated the best of the eight transistor sets tested by *Consumer Reports* in May of 1956, but it cost $75. This radio was very popular with wealthier, mobile Americans; in *Holiday* of June 1957 the set was included among the accessories needed for a picnic.

Though transistor radios were improving rapidly, if one wanted a portable that reproduced voice and music with reasonable fidelity, the only choice was a larger set with tubes. In their *Annual Bulletin* of 1956–1957 *Consumers' Research* put it bluntly:

Despite the claims that the new transistor circuits afford superb performance, better than with vacuum tubes, their performance was not found to be good. *The fact that something is new does not mean that it is better; it may often be a step backwards except in some limited particular*

13.5   Ads for some U.S.–made transistor portables of near shirt-pocket size. *Left,* Westinghouse 617P7, 1957; *right,* Admiral, 1958

characteristic. Bear this in mind in reading the ads! Transistors have two major advantages: small size and low power (battery) requirements. Used in radio they have yet to give tonal quality as good as vacuum tubes.

However, one transistor radio of this time was an outstanding performer: Zenith's redesigned Trans-Oceanic. Introduced late in 1957, the Royal 1000 Trans-Oceanic had eight bands and cost $250. (Actually, Magnavox was the first U.S. company to market a multi-band transistor portable [Fig. 13.6].) McDonald insisted that the set be compact, and it was—by Trans-Oceanic standards. But it made lavish use of metal, even for the chassis, and weighed thirteen pounds with batteries. With its nine D cells (available anywhere) the set could play for about 170 hours before needing a battery change. RCA, Philco, and other companies brought out multi-band sets, but most paled—in performance and price—next to the Zenith.

Most companies concentrated on transistor lunch box sets, which on the outside were sometimes indistinguishable from tube portables. The larger transistor radios were improving rapidly; so much so that by July 1958 *Consumer Reports* was able to state that "generally speaking, the transistor sets did better than the tube sets in CU's two most important tests—speech intelligibility under noisy conditions, and tone quality in quiet surroundings." Twenty-five portables (excluding pocket sets) were tested, both tube and transistor models. Seven models were rated best in

13.6   Magnavox multi-band transistor portable, 1957

overall quality; five of them—including the top three—were transistor radios.

By and large, transistor radios still cost more: $38 to $80 compared to $25 to $50 for tube sets. Offsetting the high initial outlay, of course, were cheaper batteries and less frequent battery replacement. Based on laboratory tests, *Consumer Reports* calculated the operating cost of each radio. The contrast was dramatic: 8 cents to 20 cents per hour for tube sets compared to 0.5 cent to 1.75 cents per hour for transistor models. Because transistor portables were ten to one hundred times cheaper to operate, their higher purchase price would be quickly repaid.

Electronic tinkerers could buy less expensive transistor portables as kits. The entrepreneurial fringe jumped in early; TraDyne (Fig. 13.7, top) and Eastern Audio, for example, began in 1955 to sell shirt-pocket radio kits. They were joined quickly by established kit makers, such as Lafayette. The latter company's KT68 and KT80 models sold for $13.75 and $6.54, respectively. Among the kit offerings from 1957 to 1958 were the Lafayette KT-116 (three-transistor) at $16.95 plus $3.75 for a "Super Power Dynamic Earphone" and the KT-119 (six-transistor) for $29.55 plus assorted accessories. Heath made several lunch-box transistor kits in the late fifties: the XR-1 had six transistors and came in leather ($34.95) and metal ($29.95) versions (Fig. 13.7, center and bottom).

The cheapest transistor radios had few transistors (one to three), and so were suitable only for young children (Fig. 13.8a–c). The Bell Products "Hit Parade" (Fig. 13.8b) contained one transistor and sold in 1958 for $8.95. A few companies also jumped in with crystal sets that resembled transistor shirt-pocket portables (Fig. 13.8d)—a design that began, ironically, with Western's post-war pocket crystal sets. A humorless *Consumer Reports* in 1959 issued several warnings to the unwary about these transistorlike radios.

Although operating costs for most transistor sets had become almost trivial, new technologies were employed to reduce such costs still further. During the mid-fifties, nickel-cadmium rechargeable batteries became available for consumer products, and a portable radio—the G.E. P715 (Fig. 13.4v)—in 1956 was the first to use them. The set came with a charger built into a travel case. Emerson also offered an optional ($20) battery charger with several models. The use of nickel-cadmium batteries and chargers never caught on; recharging had to be done frequently and was a nuisance, a task too easily forgotten.

With their very low power consumption, transistor radios could even be operated on solar cells. Admiral was the first—in mid-1956—to bring out a solar-powered transistor portable (Fig. 13.9); in fact, it was the first consumer product to be powered by a photo-voltaic device. The high-efficiency solar cells had been developed at Bell Labs in 1954, an outgrowth of transistor research. Although the radio itself was priced reasonably (at $59.95), the "Sun Power Pak" cost $175. Curiously, on the front of the plastic case was imprinted the symbol for the peaceful use of *nuclear* energy.

Another early solar model was the Hoffman "Trans-solar" CP70C ($75), which *Consumer Reports* examined in late 1958. After noting that the set "performed in a manner typical of most pocket-size transistor

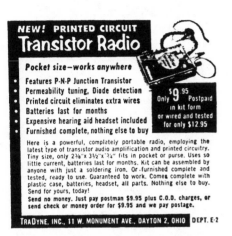

13.7 Several transistor radio kits. *Top,* TraDyne shirt-pocket set, 1955; *center,* Heathkit XR-1I, 1958 (7" high); *bottom,* Heathkit XR-1P, 1958 (7-1/4" high)

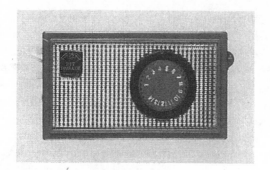

a

b

c

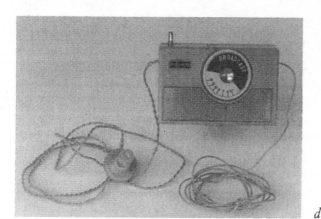

d

13.8 Small transistor radios (with few transistors) and a transistor radio look-alike. *a,* Regency XR-2, 1958; *b,* Hit Parade, 1958; *c,* Union, ca. 1958; *d,* Aud-Ion Crystal Set, ca. 1959

13.9 Admiral 7L16, solar-powered transistor portable, 1956

sets—that is, just passable," the article went on to say that its high initial price would not be quickly paid back by operation on sun power. However, they did call attention to another potential benefit: "The set does have promising possibilities, too, as an attention-getter: hold your hand over the solar cells, the music stops; remove your hand, the music is heard again. The attractive young person two seats away is watching you, obviously puzzled. Smiling, you move over to offer an explanation of this newest wonder of science." Being expensive, solar sets found little market beyond a few trendsetters.

When *Consumer Reports* published its next survey of portable radios in July 1959, no tube models were tested. Fewer than five years after the appearance of the Regency TR-1, transistors had clearly "become dominant in the portable-radio field."

Transistors would also become dominant in TVs as well. In the late fifties, major television makers had created prototype, transistorized portable TVs, but, still smarting from the color TV fiasco, they hesitated to introduce an expensive product that would appeal only to the "Cadillac trade." Even so, Philco took the plunge in June 1959 when its Safari model debuted—the world's first transistorized, portable television. The Safari had a two-inch vertically mounted picture tube that was optically magnified. Packaged in an integral leather case with a plastic visor, the Safari listed at $250. Though a technological landmark, the set sold poorly. And no wonder. The viewer had to sit directly in front, with no deviation permitted, or the picture would disappear. As a result, only one person at a time could watch TV. The Japanese would soon sell cheaper and more practical transistorized portable TVs.

The transistor radio appeared at a propitious time for the radio industry. After a huge spurt in 1946 and 1947, sales of table-model radios began to slump badly. By 1951, these sets were selling below late-thirties levels. Television, of course, had marginalized the table radio; no longer did people need new models to show off, since most sets were used in kitchens, bedrooms, and workshops. Table models with clocks kept the genre alive, but just barely. TVs, too, were in trouble. In the late fifties, set makers discovered that most families would not replace their televisions every two years; regardless of new gimmicks like remote control, the black-and-white TV market was going to be sluggish. And almost no one was buying color TVs yet. *Fortune* magazine put this into perspective in 1957: "TV has become a decidedly unglamorous area of a glamorous industry." This would not be remedied until the early sixties when color finally took off.

In 1953, almost eight million stand-alone home radios were sold; of these, about 23 percent were portables. By 1959, on sales of ten million U.S.–made sets, the proportion of portables—nearly all transistor—had reached 41 percent. Thus, transistor radios had become a profitable growth market for the industry.

In ads for early transistor radios (all of which were portables), manufacturers announced a breakthrough: because transistors did not burn out like tubes, these sets would never need servicing. For example, in early ads the Regency TR-1 was said to be "shock-resistant, virtually service-free . . . engineered for life-time performance." Westinghouse's predictions for its 1957 pocket portable (Fig. 13.5, left), in a *Holiday* ad, were no less enthusiastic: "Powered by 7 miracle transistors and the Westinghouse Silver Safeguard Chassis, it's immune to heat, vibration, moisture—*can't* wear out or burn out."

With the benefit of hindsight, these hopes for the longevity of transistors and transistor radios seem rather vain. Transistors did (and do) fail, and many of the "unbreakable plastic" cases shattered. Even if the case survived, a dropped radio usually died from a cracked circuit board. As a result, transistor radios, like their tube counterparts, sometimes needed repair. In an ironic twist, however, transistor radios turned out to be largely service-free, but only because—in sharp contrast to tube portables—they were largely unserviceable.

When a transistor radio croaked, the owner usually had no choice but to take it to a repair shop. No sane person would have contemplated taking a transistor radio apart, testing its major components, and effecting a repair. To most Americans, a transistor radio was (and is) a magical box, powered by inexpensive batteries, whose works are as inaccessible to understanding as Maxwell's invisible waves.

It was not always that way with consumer electronics. During the 1950s, more than twenty thousand supermarkets and drug stores conspicuously displayed an odd device known as a "tube checker." About the size of a refrigerator, perhaps a little narrower, the tube checker contained replacement tubes and instructions for testing all the common varieties found in radios and TVs. The amateur electronic technician usually arrived at the checker, tubes jingling together in a paper bag, children in tow, and set to work. First, it was necessary to look up the tube's number on a master roll chart having a seemingly endless roll. Once the tube was finally found, the chart disclosed the crucial information: the dial settings and which socket to use. After the tube was plugged into the correct socket, one waited for a time—a seeming eternity—to see if it registered in the "good" range. (A few sharp taps might help.) A bad tube was the most common cause of a malfunctioning radio and so the ability to find and replace a defective tube saved a technician's labor charges.

Probably few people really knew how a tube radio worked, but at least the technology had replaceable components that could be easily grasped by a human hand. Transistor radios (and later TVs) changed all that; supermarkets today do not have transistor checkers.

Although in some portables of the late fifties and sixties transistors were plugged into sockets (like tubes), mostly they were soldered in—usually for good. The owner was not the only one prevented from checking the transistors; so too were service technicians. In theory, a transistor could be tested while still soldered in, but in practice close spacing of components sometimes caused the technician to touch the wrong wire, burning out the transistor. Even if components could be checked, their replacement was difficult and time-consuming. Not surprisingly, technicians often refused to work on transistor sets because labor costs escalated quickly beyond the purchase price of a new radio; a few screaming customers usually dampened a technician's enthusiasm for tackling these

13.10 Pocket radios with tubes from the far entrepreneurial fringe, 1954–55. *Top to bottom,* Transi-Mite Micro, 1955; Western Radio Deluxe, 1955; Tinytone, 1955; Precision, 1954

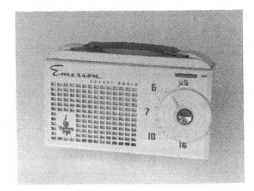

b

c

a

13.11 Tube-Transistor Hybrid Sets, 1955–56. *a*, Automatic Tom Thumb 600, 1956 (5-7/8″ wide); *b*, Emerson 839, 1955 (6-1/8″ wide); *c*, Emerson 856, 1956 (6-1/8″ wide)

high-tech radios. Most "broken" transistor sets found their way, finally, to the trash, and that is why so few early examples have survived. However, there is a gray Regency TR-1 on display at the Smithsonian Institution.

The prospective portable buyer who was unwilling to plunge into the age of throw-away electronics could for a few years still buy a radio with tubes. Though the transistor's triumph over tubes—in portables—was essentially complete by the end of the decade, dozens of tube models were built between 1954 and 1962. Surprisingly, the portables made during the tube's last stand have, in style and in electronics, a few interesting features.

The entrepreneurial fringe, including Western Manufacturing, was still trying to lure hobbyists and tinkerers with an assortment of pocket radios containing tubes (Fig. 13.10). These sets are noteworthy mostly for their size and colorful advertising copy.

A few firms on the fringe took advantage of subminiature tubes and brought out shirt-pocket FM radios. These were relatively inexpensive novelty sets that, needless to say, set no standards. Typical was the Hastings FM, Jr., of 1958. Like the earlier Privat-ear, this two-tube set had an earphone and telescoping antenna and lacked a speaker. Though cheap ($21.95 with batteries), the Hastings FM, Jr., was a flop—not because it was a mediocre performer but because it was FM only. In 1958, FM was for highbrows; listening to classical music through an ear plug was unthinkable. Not until the late sixties would rock and roll be heard on an FM station.

Among the most fascinating sets from the early transistor era were small pocket portables with subminiature tubes *and* transistors (Fig. 13.11a–c). Emerson and Automatic converted their micro-table-models into tube/transistor hybrids. This made sense at the time (when transistors were still fairly expensive): two transistors could replace the heavy-drain audio output tube. Thus, the hybrid sets could be operated more economically than all-tube sets and were less expensive than all-transistor models. More importantly, perhaps, the radios could be advertised as containing transistors, thus exploiting that high-tech cachet.

Indeed, both the Emerson 838 and Automatic "Tom Thumb" 600 displayed the word "transistor" prominently on the front of the case. Curiously, in one later version of the 838 (Model 856), Emerson removed the transistor designation, perhaps responding to industry criticism that it was misleading the public (after all, the radio still had three subminiature tubes).

The most unusual hybrid set was the Crosley JM-8 of 1955 (Fig. 13.12). Like the others, it contained three subminiature tubes and two transistors, but it was far from being a recycled design. The Crosley JM-8 looked like a book, including leather binding. Six versions were available, differing in color and title, at $50 each. The JM-8's operating cost was a not noteworthy eleven cents an hour, though it played better than all-transistor sets selling for $50. The JM-8 was one of the last models to carry the brand of this old line radio company (which had been absorbed into AVCO).

Most of the remaining tube portables made during the transistor age conform to the lunch-box style (Fig. 13.13). Even so, different materials and colors allowed for considerable variety. Motorola offered tube portables in metal cases covered with "scuff proof, stain-resistant miracle fabric" in various colors, with their fronts decorated in plastic or chrome (Fig. 13.14a–f). Even today, these radios—which cost between $30 and $50—have a pleasing appearance. Motorola's Model 56L was available in charcoal, pink, flame red, and cerulean blue, colors similar to those of cars built at the same time (1956). Looking at the radios today, one cannot help but chuckle. On a few models, clever Motorola designers had configured the knobs and grill to look like a face. At the sales counter, who could resist a smiling radio? (Even Motorola's earlier Model 53L—Fig. 10.9k—resembled a cute robot's head.)

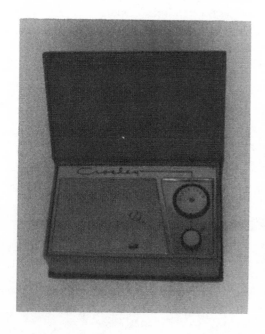
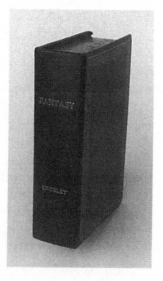

13.12 Two views of the Crosley JM-8, 1955. *Left,* open; *right,* closed (7″ high)

An all-metal case required that the antenna migrate outward, and so it was placed in the plastic handle. Turning a necessity into a virtue, Motorola ads called attention to its "Exclusive Roto-tenna handle": "You just turn the handle (not the radio) for stronger, clearer reception." The swiveling antenna actually worked pretty well and was copied by other companies—even on all-plastic radios. RCA, for example, made several models having swivel antennas just beneath the handle (Fig. 13.13m, n, and r). In *Popular Science Monthly*'s review of portable radios in August of 1957, rotatable antennas were regarded as a desirable feature and their presence or absence was noted (eight of twenty-seven tube radios had them).

The leather look revived with vigor. Not content with simulated leather fabric or plastic mottled to resemble leather, radio makers used the real thing for the entire case. On television, westerns like "Gunsmoke" and "Wyatt Earp" were much in vogue during the 1950s, and leather portables evoked the cowboy image. Philco's Model 655 was typical in its "Top Grain Cowhide Case." This piece of the great outdoors could be had for only $39.95, including cowhide aroma (Fig. 13.15).

The most bizarre leather lunch-box portable of this time was the Sylvania U-235, which sported a compass, sundial, and Geiger counter. Why would someone want to measure radiation levels with a portable radio? This was the fifties, of course, the height of Cold War madness, when as a weekend pastime some families built a backyard bomb shelter. In fact, Civil Defense literature listed portable radios among necessary bomb-shelter accoutrements. For this purpose, the Sylvania U-235 was just perfect: after the nuclear holocaust, survivors could seek signs of civilization (on the radio) and periodically test radiation levels outside. Once it was safe to venture above ground, the compass and sundial would come in handy. Happily, this radio—which cynically exploited Americans' darkest fears—did not spawn a new genre of portables.

Plastic lunch boxes were still very popular (Fig. 13.13), and most companies offered one or two models. For the first time, some plastic cases bear the imprint "unbreakable." However, Father Time (and carelessness) have taken their toll; today, many of these unbreakable cases are broken. As a group, these radios are undistinguished; the golden age of plastic lunch-box portables had already passed. A few sets had metal trim, and some a second plastic color, but even these lack a unity of style to catch—and hold—the eye. It is easy to suspect that the industry's best designers were now working on other products, like television.

The largest tube portables of the day, whose styles had been frozen in the forties, were multi-band sets, such as the Zenith Trans-Oceanic. Surprisingly, the Trans-Oceanic with tubes (Model B600) was still being made in 1962—the last tube portable manufactured in the United States (Fig. 13.16). Efforts to reduce production costs of tube portables led, beginning in the middle fifties, to the adoption of printed circuit boards and integrated circuits. Many of the last tube portables had these features. Printed circuit boards appeared in portables by at least 1954, with Motorola's 54L (Fig. 13.17, top) one of the first. In addition to Centralab's integrated circuits were "tinkertoy modules"—integrated

a

b

c

d

e

f

g

h

i

j

k

l

m

n

o

p

q

r

s

t

13.13 An assemblage of tube portables of
the transistor age, 1955–58. *a*, RCA 6BX41B,
1955 (10-1/8″ wide); *b*, Westinghouse H495P4,
1955 (9″ wide); *c*, General Electric 661 (radio)
and 661A (clock and power supply), 1955
(13-1/4″ wide); *d*, Bulova Companion 204, ca.
1955 (9-1/8″ wide); *e*, RCA 7BX5J, 1955 (6-1/2″
high); *f*, General Electric 637, 1955 (9-3/4″
wide); *g*, Admiral 217, 1958 (10-3/8″ wide); *h*,
Zenith Santa Monica, ca. 1955 (7-1/2″ high);
i, Roland Riviera 4P2-2, 1956 (6-1/4″ wide); *j*,
Emerson 850, 1956 (8″ wide); *k*, Airline GEN
1103A, 1956 (8-3/8″ wide); *l*, Emerson 879B,
1957 (6-5/8″ high); *m*, RCA 8BX-6L, 1956
(10-1/2″ wide); *n*, RCA 7BX8J, 1956 (11″ wide);
o, Sentinel 1U-355P, 1956 (7″ high); *p*, Zenith
A400, 1957 (10-1/4″ wide); *q*, Zenith 5Y40,
1956 (11-1/8″ wide); *r*, RCA 7BX6E, 1956
(10-1/2″ wide); *s*, Zenith A402, 1957 (10-1/2″
wide); *t*, Arvin 8571-1, 1958 (11″ wide); *u*, Arvin 8572, 1958 (11″ wide); *v*, Westinghouse
H662P4, 1958 (7-5/8″ wide); *w*, General Electric P672, 1958 (6-3/4″ high); *x*, Motorola
5P21R, 1958 (10-1/2″ wide)

u

v

w

x

13.14   Motorola tube portables in metal
cases, 1955–57 (10-1/4" wide). *a,* 55B1, 1955;
b, 55L3, 1955; *c,* 56L4, 1956; *d,* 56M3, 1956; *e,*
56B1, 1956; *f,* 5P33VW, 1957

13.15   Philco 655 tube portable, 1955

13.16 Zenith Trans-Oceanic B600, the last U.S.–made tube portable

13.17 Motorola printed circuit chassis. *Top,* 54L, 1954; *bottom,* 5P32, 1957 (tinkertoy module is below elevated tube at center)

resistor-capacitor networks into which tubes could be plugged. Tinkertoy modules were developed in the early fifties by the National Bureau of Standards for the military, but they soon entered portable radios. Motorola, among other set makers, used them extensively in the mid-fifties (Fig. 13.17, bottom). Though more expensive than the individual components they replaced, integrated circuits reduced handling and wiring costs. Like printed circuit boards, integrated circuits were a serviceman's nightmare, being difficult to test and replace, but they were very reliable.

Motorola's last tube portables are especially noteworthy because of how much labor had been squeezed out of the assembly process. These sets employed printed circuit boards and varying combinations of tinkertoy and Centralab integrated circuits. Fewer discrete components were used in these portables than in any superheterodyne radio built (with tubes or transistors) until the 1980s. Moreover, soldering was done mechanically, in a sophisticated dipping operation. These technological innovations, it must be stressed, were adopted to reduce labor costs, not to promote miniaturization (after all, they were lunch-box portables). But they also resulted in more reliable radios.

Because national advertising for tube portables had all but ceased in 1958, at first glance it is surprising that any were being sold as late as 1962. Yet, compared to lunch-box transistor sets, tube portables were still cheaper, making them the radio of choice for people on a tight budget. In fact, tube portables were sold as loss leaders by discount houses simply to bring in buyers. More importantly, gift-givers, who still accounted for a good number of portable sales, in many cases would have been less concerned with the higher maintenance costs than the recipients. Their bottom line was *purchase price.* Finally, a few aficionados earnestly believed that tubes sounded better—more mellow—than transistors. In any event, the transistor's triumph was rapid.

By 1960 the tube portable was all but extinct. Though transistor radios were costly or impossible to repair and sounded no better than tube sets, their marvelous battery economy proved decisive. As the cost of transistors—and thus transistor sets—came down, there was no contest. The transistor was high-tech and cheap to operate; tubes had become the relic of an earlier age. Tubes would survive longer in other electronic products (until the late 1960s in stereos and table-model radios and the middle 1970s in televisions), but the overall outcome was never in doubt. The transistor revolution, so brashly forecast in 1948 by Bell Labs, had a decade later begun in earnest.

14

Made in Japan

AS MANY A MODERN COMMENTATOR HAS OBSERVED, IF A latter-day Rip Van Winkle had nodded off in 1944 and awakened today, he would doubtless conclude that the Second World War had been won by Japan. The evidence, after all, is abundant and compelling: Toyota, Nissan, and Honda cars carry us to work and to play; Nikon, Canon, and Olympus cameras capture our favorite scenes on celluloid; and Panasonic, Sanyo, and SONY electronic gizmos infuse our homes with sights and sounds. Though Japan did lose on the battlefield, in the marketplace American companies have been retreating steadily for decades.

The first product to be totally overwhelmed by Japanese competition was the shirt-pocket portable. This unassuming radio, bought to play rock and roll, launched Japan on a course toward world leadership in consumer electronics. Surprisingly, this rise to preeminence began before Japan had become a consumer society.

Japan, an island nation, is about the size of Montana. Though deficient in natural resources, this tiny bit of real estate now supports in excess of 120 million people. Less than 20 percent of the land can be farmed, so even food must be imported.

When Meiji leaders in 1868 put the nation—still a feudal society—on a path toward Western industrialization, they were aware that, to pay for imported raw materials, fuels, and capital equipment, Japan would have to develop products for export. And she did. By the 1930s, when Japan was governed by the military (allied with large trading and industrial firms), the country had inched up behind the United States, the United Kingdom, and Germany to become the fourth largest exporting nation. Textiles, pottery, canned fish, raw silk, and tea, for example, were carried, usually in Japanese-made ships, to ports around the world. In the mid-thirties, the United States bought 17 percent of Japan's exports.

Though Japan had become an industrial power, most of her people were still peasants. In pre-war years, consumer goods common in the United States were luxury items owned by only a handful of well-heeled Japanese. Fewer than one home in a hundred had a refrigerator, vacuum cleaner, or electric washing machine. Most houses lacked central heating and indoor plumbing, and were too tiny to accommodate Western-style furnishings. Passenger cars, of course, were an oddity. Needless to say, one could not buy consumer goods on the installment plan. Indeed, well into the sixties, to be in debt was regarded as a disgrace. Japan had industrialized, but its export goods and military products did not raise the standard of living of ordinary people. However, ordinary people—long accustomed to subordinating individual (even family) desires to a greater good—accepted impoverishment of the consumer sector, believing that the nation's survival depended on hard work, sacrifice, and thrift to maintain the flow of exports.

The Second World War left the Japanese economy in shambles. Her colonies stripped away, her industry in ruins, and her people hungry, Japan had to make a new beginning. But Japan now had a wealthy and generous friend: the United States. As a matter of fact, American Far Eastern strategy was keyed to having a strong, closely allied Japan. The Occupation Forces, under the command of General Douglas MacArthur, gently imposed a Western-style democracy and new constitution (though leaving the emperor intact), and dispensed $2 billion in American aid. Most importantly, the United States mandated that no military-industrial complex could again take control. Japan would have to rebuild her industrial base on a nonmilitary basis.

In the next decade the Japanese, with American assistance, forged a new economy and new society. Without its military and captive colonial markets, industry looked ahead to the consumer sector. However, the first priority was to refurbish the industrial infrastructure. Thus, the decision was made to starve consumers a while longer, at least until the anticipated demand could be met by *Japanese* firms. Heavy industries—iron, steel, ship-building, chemicals, electrical, and so on—were rebuilt, and the yen was undervalued to make Japanese goods attractive abroad. During the Korean War, the United States placed $4 billion in orders with Japanese companies, a significant priming of the industrial pump.

Japanese goods exported to the United States in the late forties and fifties had an unsavory reputation. To Americans, "made in Japan" simply meant junk—"Jap crap." Indeed, some U.S. officials and foreign observers doubted that Japan could make products that would appeal to Americans, but this concern was soon shown to be unfounded. Beginning in the fifties, a few Japanese cameras, telescopes, binoculars, and sewing machines were sold in the United States. The quality of such products initially was uneven, but soon Japan began to monitor exports. American attitudes toward Japanese products would change decisively in the sixties; Japanese companies would be looked upon as champions of the U.S. consumer.

Although post-war Japan was ostensibly a democracy, in fact the nation quickly evolved into a technocracy with centralized planning and management. Economic policy was made by government officials—many of them respected engineers—with close ties to industry. This core of competent professionals was insulated from the whims of public opinion and the pressures of special-interest and consumer groups. In this unique form of organization, which still largely prevails, elected officials are mostly window dressing.

In Japan, the real power resides in two government agencies: the Ministry of Finance and the Ministry of International Trade and Industry (MITI), which have been single-mindedly dedicated to promoting economic growth and exports. Among its many duties, MITI decides which industries will be targeted for future growth, which companies will have access to foreign currency and technology, and which goods will be protected by tariffs and red tape from imports. In addition to maintaining and refining Japan's export orientation in the fifties, MITI targeted a few consumer sectors for nourishment. One of these was electronics. With ample incentives, Japanese companies began to produce radios and TVs, and to test the export waters.

a

The first Japanese electronic product to be marketed widely in the United States was small portable radios with tubes. In the mid-fifties (especially 1956 and 1957) hundreds of thousands of these tube portables found their way into American homes (Fig. 14.1). Though occasioning little comment at the time by consumer magazines or U.S. radio companies, these sets are of interest today for what they reveal about Japanese radio technology before the transistor. (Some later radios are also shown in Fig. 14.1.)

How many companies were actually producing tube portables at this time is anyone's guess. It is clear, though, that there were far fewer companies than radio brands. In a strategy that would be pursued vigorously with transistor portables, some firms marketed the identical radio under different brands (Fig. 14.1j, k).

b

A few unremarkable sets, apparent mimics of the RCA B411, were tiny plastic lunch boxes. However, many tube portables were quite small (a few even shirt-pocket size), indicating that Japanese radio makers were already emphasizing miniaturization. Some sets were obvious copies of the Emerson 747 and Automatic "Tom Thumb" 528 (the smallest U.S. superheterodynes with tubes from the mid-fifties). Subminiature tubes were used rarely in these clones, but the Excel KR-451 is one example (Fig. 14.1c); the subminiature tubes were made in Japan without benefit of a Raytheon license. This petite radio measured 5-3/8" × 3-1/2" × 1-5/8", and had a leather case (with zipper to allow access to its insides). Technologically, the Excel KR-451 contains no surprises; extreme compactness was achieved simply by crowding the parts on a thin metal chassis.

c

The vast majority of pocket portables used miniature tubes, which were made in Japan under license to RCA. Apparently unconcerned with tone quality and battery economy, Japanese manufacturers compressed these sets as never before. Among the smallest was the Little Pal (Fig. 14.1i); at 5-3/4" × 3-1/8" × 1-3/4", perhaps the most diminutive superheterodyne ever made with four miniature tubes. Miniaturization was effected by use of a tiny B battery and close packing of components. Like most pocket portables with tubes, Little Pal would have gobbled up batteries at a prodigious pace.

A few manufacturers came up with a clever design that improved battery economy while enhancing miniaturization. The Star-Lite RN-4B, for example (Fig. 14.1d), was a superheterodyne with just three tubes. By dropping the final audio stage, engineers eliminated not only a power-

14.1 Japanese tube portables (these radios, which sometimes lack model numbers, are difficult to date). *a*, Harpers GK-301, ca. 1960 (7″ wide); *b*, Excel T311, ca. late fifties (6-1/4″ wide); *c*, Excel KR-451, ca. 1956 (5-1/8″ wide); *d*, exterior and interior views of Star-Lite RN-4B, 1956 (5-1/8″ wide); *e*, Sterling PR-537, ca. 1960 (6-1/2″ wide); *f*, Stantex B-33, ca. late fifties (7-1/8″ wide); *g*, Comet, ca. late fifties (6-1/2″ wide); *h*, Comet, ca. late fifties (6-3/8″ wide); *i*, Little Pal L-400, ca. late fifties (5-7/8″ wide); *j*, Silver (Shirasuna Denki), ca. 1956 (5-1/8″ wide); *k*, Stantex (Shirasuna Denki), ca. 1956 (5-1/8″ wide); *l*, Vanguard, ca. 1961 (7-3/8″ wide)

hungry tube but also the speaker. Clearly, there was no free lunch: listening was through an ear plug. Surprisingly, the set played pretty well, and the life of the A battery would have been extended appreciably. This tiny radio (measuring 5-5/8″ × 3-1/4″ × 1-5/8″) was sold during the summer of 1957 by U.S. discount houses for $10. Radios of similar design were marketed in 1956 and 1957 for as much as $20. Because the Japanese tube portables were smaller than the U.S. sets that had inspired them but were otherwise undistinguished, it is tempting to dismiss them as little more than technological one-upmanship. Yet, there were long-standing incentives for Japanese manufacturers (of radios and most other products) to miniaturize as much as possible. Smaller radios—with small parts inside—meant reduced use of imported materials for making components, chassis, and case. In addition, lighter, more compact radios could be handled and shipped abroad more cheaply. With the coming of transistor technology, Japanese companies would miniaturize portable radios with unprecedented vigor.

The first Japanese transistor radio was marketed in the United States—in 1957—by SONY. This firm's beginnings go back to 1945, and they were not auspicious. In the ruins of a bombed-out department store in downtown Tokyo, entrepreneur Masaru Ibuka founded a small company, beginning with seven employees in one room. Together with Akio Morita, a physicist by training, Ibuka struggled to create a company that could turn the latest scientific and engineering advances into distinctive products for daily life. The similarity of this vision to American consumerism was no coincidence. American products had been eagerly embraced before the war by a Westernized Japanese elite. Heir to a *sake* fortune, Morita recalled his comfortable childhood, surrounded by American things:

> I remember Sunday outings, riding in an open Model T or Model A Ford, bumping along the rutted, narrow and dusty roads at a very slow speed, my mother sitting in the back seat in a very dignified and stately way holding her parasol upright to shade her from the sun. Later, father used to go to work in his chauffeur-driven Buick. At home we had a General Electric washing machine and a Westinghouse refrigerator.

Their new company, christened Tokyo Telecommunications Engineering Co., was dedicated to promoting a Japanese version of consumerism.

The first product of Tokyo Telecommunications was an electric rice cooker, which seemed like a good idea at the time because fuel was still scarce. It did not work well; and not one of the hundred made was sold. The fledgling company survived the tough times by manufacturing specialized electronic products for the government and industry. Their first major success was a tape recorder. Building a working machine, based on German technology, was not nearly as difficult as making the tape; after much trial and error, they succeeded on both counts. Improved versions of the tape recorder made inroads into Japanese courts and schools in the early fifties. In a strong patent position, Tokyo Telecommunications held a virtual monopoly on tape recorders in Japan, profits from which financed the company's other electronic ventures.

In 1952, Ibuka went to the United States for three months to find new uses for tape recorders. While there, he learned that AT&T would soon start licensing the transistor patent to all comers who could cough up a $25,000 advance on royalties. It was obvious to Ibuka, as it had been to most Americans in the industry, that this bizarre bit of crystal held the future of electronics. But how could his company use it? An idea quickly took shape: the transistor would provide a stimulating challenge to his young engineers. Indeed, Morita and the others back in Tokyo were enthusiastic.

Before Tokyo Telecommunications could set to work, MITI had to be convinced that paying AT&T's sizable fee was in Japan's best interests. In Japanese industry, Tokyo Telecommunications was on the entrepreneurial fringe, not apt to receive MITI's favors; but it did.

Having hurdled the bureaucratic barriers, Tokyo Telecommunications had to select a product. Japanese engineers were voracious readers of American electronic literature and were well aware of the devices hobbyists and tinkerers were building with transistors. After surveying the possibilities, they chose to make a radio. Not just any radio, of course, but, in Morita's words, "a radio small enough to fit in a shirt pocket." It appears that cultural imperatives sometimes move easily from one culture to another. But Ibuka, Morita, and the others were technical people who shared the visions of this group, which transcends ethnic and national boundaries. The Japanese entrepreneurs apparently had not heard about the Belmont Boulevard.

The first task at hand was to make transistors. Like American companies at the same time, Tokyo Telecommunications found the going rough. To keep abreast of American progress, Ibuka and another company official in early 1954 visited U.S. laboratories and factories engaged in transistor work. At the end of each day on the three-month trip, Ibuka and his colleague dispatched a detailed report back to Tokyo where experiments were in progress. By June 1954 Tokyo Telecommunications had made their first functioning transistor.

The Japanese transistor work was not mere imitation, as many Americans believe. Much original research eventually was done to improve the transistor's high-frequency performance and power-handling capability, and new kinds of transistors were created. Years later, Leo Esaki was awarded a Nobel prize for research he carried out at SONY in the mid-fifties.

After the transistor problem was solved, engineers at Tokyo Telecommunications had to design new, miniaturized components for their tiny radio. The production of these components was in some cases farmed out to other companies, thereby stimulating an entirely new industry.

Tokyo Telecommunications was finally ready to launch its transistor radio. But first things first: a clever, catchy brand name was needed. After much research and debate, they selected SONY. Their transistor radio, model TR-55, was put on the Japanese market in August 1955— almost a year after the debut of the Regency TR-1—and was the first product to wear the new SONY brand. In January of 1957, Tokyo Telecommunications changed its name to SONY.

The SONY TR-55 was a small transistor portable but, being about the same size and format as the Emerson 747, was beyond the capacity of shirt pockets. Further miniaturization was ordered. In March of 1957, the TR-63 was produced and marketed in Japan, SONY's first shirt-pocket portable (Fig. 14.2). Though slightly smaller than the Regency TR-1, the set was a little larger than Japanese shirt pockets of that time. This would not do. Shrinking the radio further was not immediately feasible, so SONY decided instead to expand the shirt pocket. A special batch of shirts was issued to SONY salesmen, who could then demonstrate to prospective clients what they erroneously believed was "the world's first pocket-size all-transistor radio."

Inside, the SONY TR-63 was a fascinating mix of old and new; yet, its technology clearly initiated trends that later Japanese shirt-pocket sets would follow. Like most U.S. portables of 1957 (tube and transistor), the SONY set had a printed circuit board. However, its design was somewhat less than elegant, and required labor-intensive fixes, including the use of wires and the soldering of a half dozen components on the board's underside. In addition, all connections were soldered individually by hand. Evidently, SONY could employ cheap labor instead of high-tech manufacturing operations. In fact, Japanese labor costs in the late fifties were about one-seventh of those in the United States.

The SONY TR-63 contained a tuning capacitor of innovative design that was breathtakingly tiny. Such tuning capacitors would be adopted almost universally by Japanese companies for their shirt-pocket portables. This radio also introduced a new nine-volt battery that became standard in small transistor radios.

On the eve of SONY's triumph (marketing the first Japanese transistor radio in the United States), Japan's progress toward a consumer society was still limited. A land of ninety million people in 1956, Japan had only one million TVs. Radio production was rising, however—to three million sets in 1956. In that year most American homes had two or three radios; most Japanese homes still had only one, probably made before the war. Passenger cars remained rare; even washing machines, refrigerators, and vacuum cleaners were uncommon. For a nation such as this, still recovering from its war wounds, to sell a high-tech consumer product to people in the world's mightiest industrial nation would be a significant coup.

In typical Japanese fashion, SONY began to seek foreign—especially U.S.—markets for its shirt-pocket radio. In the United States, SONY distributed its set through department stores and other retail outlets that could showcase the pricey pocket portable. At $39.95, the TR-63 cost about the same as comparable American-made sets and was not exactly a hot seller (in fact, SONY made no money in the United States for several years).

Nonetheless, SONY soon inspired a host of imitators. In 1958 and 1959, around three dozen Japanese firms began exporting small transistor portables to the United States. About 100,000 were imported in 1957; in 1959 more than six million entered the U.S.—*half of Japan's total radio production,* nearly all of them shirt-pocket transistor sets. Income earned from these sales was important: radios were in 1959 Japan's second largest source of U.S. dollars—sixty-two million of them—and

14.2   SONY TR-63, the first Japanese transistor radio sold in the United States, late 1957 (4-3/8" high)

14.3. Ultraminiature Japanese transistor radios. *a,* Toshiba 6TP-357, 1959 (2-15/16″ high); *b,* Marvel 6YR-15A ca. 1960 (3-1/2″ high); *c,* Marvel JL62, ca. 1960 (4-1/2″ wide)

their fourth largest export. MITI officials were ecstatic.

SONY and the other Japanese radio makers had lucked out: their foray into the U.S. market coincided perfectly with the proliferation of all-rock radio stations. Rock music had created millions of teenage consumers eager for these tinny, tiny sets.

Not only did small transistor portables open up the American—and ultimately a world—market for Japanese electronic goods, but it launched a new industry: consumer microelectronics. With the shirt-pocket portable, Japanese manufacturers began to cultivate the technologies that eventually would make them the world's leader in microelectronics. It is curious that the growth of Japan's microelectronics industry was catalyzed by the musical tastes of American youth.

What most distinguishes the Japanese shirt-pocket sets is their small size. Tendencies toward miniaturization, already present in Japanese tube portables, were intensified with transistor technology. Lower voltages in transistor circuits permitted a quantum leap in miniaturization, which was achieved in two ways. First, components were crowded together on the circuit board to an extreme degree. This strategy required more labor because of the tight spaces in which components had to be manipulated. Second, components themselves were shrunk further, especially intermediate-frequency transformers and speakers. Using these miniaturized components, Toshiba, beginning in 1959, made several shirt-pocket radios a hair under three inches tall (Fig. 14.3a). The Toshiba 6TP-357 sold for $29.95 and was claimed to be the "World's Smallest" six-transistor radio. The ad exulted: "Styled like an artistic miniature, precision engineered like a precious wristwatch, a marvel in electronic engineering." Other Japanese manufacturers also considered their miniature sets to be marvels, and one (Yokohama Tsushin Kogyo) even adopted Marvel as the brand name (Fig. 14.3b, c).

To American radio makers, the shrinking derby made no sense for consumer products; smaller portables would simply sound tinnier—it was technological virtuosity for its own sake. Yet, because their Lilliputian sets enjoyed robust sales in the United States, the Japanese could only conclude that Americans *loved* miniaturized electronic products.

It is apparent that Japanese firms were scaling down all components more or less to transistor size (perhaps taking their cue from American military electronics). To the Japanese it was obvious that a world-class electronics industry for the transistor age would require very tiny components.

Unlike the miniature parts produced by U.S. suppliers for Cold War hardware, Japanese components were made by cheap labor and sold at low prices. As a result, Japan soon had an enormous capacity to produce inexpensive transistors and transistor-compatible components of high quality. Though miniaturized components did not improve radio performance, they did serve as the foundation of Japan's entire consumer electronics industry as it converted from tubes to transistors. Japan underwent this conversion quickly because transistorized products have vastly lower power consumption—which mattered greatly in a country that imported nearly all of its energy.

Consumer magazines began to test Japanese transistor portables in

n

o

p

14.4 Some Japanese transistor pocket portables of the late fifties and early sixties. *a,* Hitachi TH-621, ca. 1958 (4-1/2″ wide); *b,* Channel Master 6506, 1959 (6″ wide); *c,* Linmark T-25, ca. 1959 (4-1/2″ high); *d,* Channel Master 6509, 1960 (3-3/4″ high); *e,* SONY TR-810, ca. 1961 (5-1/4″ wide); *f,* Jaguar 6T-250, ca. 1960, (3-7/8″ high); *g,* Mellow Tone, ca. 1961 (3-3/4″ high); *h,* Coronet, ca. 1960, with and without case (4-1/2″ high); *i,* Universal PTR-81B, ca. 1962 (4-3/8″ high); *j,* Crown TR-680, ca. early sixties (3-1/2″ high); *k,* Crown TR-400, 1960 (4-7/8″ wide); *l,* United Royal 801-T, ca. 1962 (4-1/2″ wide); *m,* Matsushita T-13, ca. 1961 (4″ high); *n,* Trancel T-11, ca. 1962 (4-1/2″ wide); *o,* Universal PTR-62B, ca. 1962 (4-3/8″ high); *p,* Raleigh, ca. early sixties (3-5/8″ high)

1959; *Consumer Bulletin* assessed eleven radios, mostly shirt-pocket size. The article noted that Japanese radios were generally less expensive than American-made equivalents, and also expressed surprise that the components and workmanship found in the imports met high standards.

As a group, the Japanese radios of the late fifties and early sixties were rather pleasing to the eye (Fig. 14.4) with their graceful lines and lively colors. Without doubt, this was the golden age of the transistor pocket portable.

Beginning in the late fifties, some American companies sounded the alarm, trying to convince federal officials and Congress that Japanese transistor radios were a threat to national security. These charges, ludicrous at face value, were soundly rebutted by the Japanese. Moreover, other American companies already were collaborating with Japanese firms, and so did not wish to join the fight and jeopardize their profitable relationships.

For various reasons, the U.S. government took no action against Japanese imports. Many powerful people staunchly opposed trade restrictions on principle. For example, Robert Galvin, president of Motorola, was a spokesman for "unrestricted trade throughout the world." Many experts believed that high U.S. tariffs in the thirties had spread and prolonged the Depression. Consequently, U.S. policy was (and still is) firmly in the free-trade camp. And, of course, the United States in the fifties had a huge trade surplus with Japan.

It was also U.S. policy to support Japan's economic recovery; a strong anti-communist ally was needed in the Far East if South Korea was to survive. Thus, Japan's success in the U.S. marketplace was welcomed. Cold Warrior Sarnoff himself, in 1960, applauded Japan's achievements in electronics, claiming that they were good for both countries.

It should also be recalled that inexpensive, high-quality products were thought to be in the best interests of consumers—regardless of their country of origin. In a consumer society, such a position carried great weight.

Though political factors were not stacked against the Japanese, in 1960 they nonetheless "voluntarily" adopted export quotas for AM radios having three or more transistors. Japanese radio firms, however, were adept at getting around these restrictions. They simply stepped up shipments of one- and two-transistor "toy" radios, multi-band radios (Fig. 14.5), and, unbelievably, portables with tubes (Fig. 14.1a, e, l). Also, using a tactic that was to become important in later battles over color TV, Japanese companies got around the agreement by shipping six-transistor sets through third countries. In addition, minimum factory prices on export radios, set by the Japanese government, were violated flagrantly and had to be adjusted downward; in 1961 they were suspended altogether.

Few U.S. companies had made shirt-pocket portables in the mid-fifties but, during 1959 and 1960, nearly all began to offer them in response to Japanese imports. Although the teenage market tapped by the Japanese was vast, profits on the tiny sets were apt to be low because of intense price competition. U.S. firms differed in the ways they coped with antici-

14.5 Japanese multi-band transistor portables (ca. 1960–64). *a,* Kobe Kogyo KT-1000, ca. early sixties (6-1/2″ wide); *b,* Hitachi WH-822H, 1963 (6″ wide); *c,* Jupiter TR-8s, ca. early sixties (6-1/8″ wide); *d,* National (Matsushita), AT-290, ca. early sixties (6-1/8″ wide); *e,* Panasonic T-100D, 1964 (13-1/8″ wide); *f,* Aiwa AR-122, 1964 (9″ wide)

pated low profits. These varied responses and the ultimate outcomes tell us much about how the United States began the retreat from consumer electronics more than three decades ago. American companies learned lessons from their head-to-head competition with the Japanese on shirt-pocket portables that they would apply later when the competition shifted to large portables, then to black-and-white and color TVs.

In 1959, several companies, including Bulova, Olympic, Magnavox, and Columbia, immediately contracted with Japanese firms to supply shirt-pocket radios that would carry their American brand names (Fig. 14.6). Surprisingly, among the companies making this move was Motorola, the U.S. leader in adopting labor-saving innovations. Motorola had apparently concluded that even their efficient manufacturing techniques could not compete with the cheap parts and cheap labor that went into Japanese sets. They saw immediately what other manufacturers would learn later.

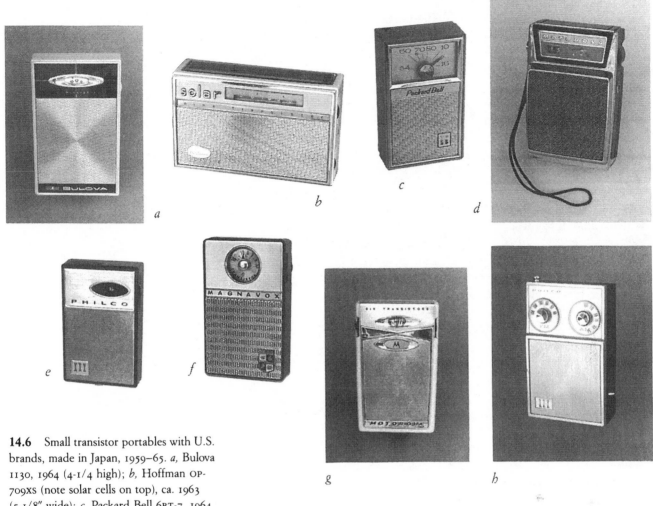

14.6 Small transistor portables with U.S. brands, made in Japan, 1959–65. *a,* Bulova 1130, 1964 (4-1/4 high); *b,* Hoffman OP-709XS (note solar cells on top), ca. 1963 (5-1/8″ wide); *c,* Packard Bell 6RT-7, 1964 (4-1/4″ high); *d,* Motorola X15E-1, 1960 (4″ high); *e,* Philco NT600BK6, ca. 1963 (3-3/4″ high); *f,* Magnavox AM-60, 1961 (4-1/4″ high); *g,* Motorola X11B, 1959 (4-1/8″ high); *h,* Philco NT814BKG, 1965 (4-1/2″ high)

Motorola's first made-in-Japan model was the X11B (Fig. 14.6g), which was doubtless built by Toshiba. The label inside the case says "printed in the U.S.A.," and there is no indication that the radio itself was made elsewhere. Typically, made-in-Japan radios with American brand names were not conspicuously identified as such.

In these collaborations, companies like Motorola transferred their technologies to Japanese firms, often receiving royalties. RCA, though still making its sets at home, eagerly licensed technologies to Japanese companies. In so doing, RCA had earnings on many radios and TVs made in Japan—wherever sold. Ironically, U.S. tax and tariff policies underwrote "offshoring" of manufacturing as well as technology transfers. In practice, though, Japanese companies adopted the American technologies they needed, with or without the necessary licenses.

Because Japanese components were so inexpensive, some companies incorporated them into their U.S.–assembled shirt-pocket sets (Fig.

14.7). With dip soldering and the cheaper foreign parts, these hybrid radios could be mass produced and offered at competitive prices—for a while.

Still other companies—RCA, Zenith, GE, Westinghouse, Admiral, and Philco—took on the Japanese with all-American radios (Fig. 14.7). An optimistic Westinghouse executive, C. J. Urban, put it like this: "We can meet foreign competition with innovation and creativity." Benjamin Abrams, Emerson's founder and president, was even more upbeat, predicting in mid-1959 that "the Japanese import will kill itself off and disappear—within the next 12 months we will see a material decline in Japanese imports." Apparently, Abrams thought that the American infatuation with inexpensive, high-quality Japanese products was a fad, about to run its course like the previous year's Hula Hoop.

Philco alone introduced three all-American shirt-pocket portables in 1959. Zenith also mounted an aggressive all-American campaign, marketing two shirt-pocket models in 1960 (the Royal 50 at $30 and the Royal 100 at $40), targeted at teenagers (Fig. 14.8).

How did the American-made shirt-pocket radios fare? Not so well. Although some all-American sets were selling in 1960 for $30, a few Japanese models had dropped to $20. The price pressure was so severe that many more companies switched to Japanese parts or Japanese manufacture of their sets.

Surprisingly, in the spring of 1962 a few American-made hybrids descended to $15, matching the still-plummeting imports. Nonetheless, it was a lost cause, especially when Japanese firms or their affiliates in Hong Kong and other East Asian nations flooded the market with even cheaper radios.

By the end of 1963, no all-American sets survived. Just a few years later, no company was making a shirt-pocket set in the U.S. Ironically, many Japanese companies also stopped producing shirt-pocket portables at home in the middle and late sixties, as the income of assembly-line workers began to rise.

14.7 Transistor shirt-pocket portables made in the United States, 1960–65 (after 1961, most contained some Japanese parts). *a,* Admiral Super, 1960 (4-1/2" high); *b,* General Electric P911J, ca. 1963 (4-5/8" high); *c,* RCA 1RJ19, 1961 (4" high); *d,* RCA 3RH34, 1961 (4" high); *e,* Arvin 61R13,23, 1961 (4-5/16" high); *f,* General Electric P850D, 1962 (3-1/2" high); *g,* General Electric P825A, ca. 1961 (3-7/8" high); *h,* General Electric P915C, 1964 (4-3/4" high); *i,* Sears Silvertone 42081, 1963 (4-1/2" high); *j,* Zenith Royal 40, 1963 (4-3/8" high); *k,* Arvin 64R38, 1964 (4-3/8" high); *l,* Admiral Y2223A, 1962 (3-3/8" high); *m,* General Electric P955B, 1965 (4-3/4" high); *n,* General Electric P1710C, 1965 (4-3/8" high); *o,* General Electric P1704A, 1965 (4-3/4" high)

14.8 The Zenith Royal 100 was marketed to teenagers (from an ad in *Senior Scholastic,* 1960)

a

b

c

d

e

f

g

h

i

j

k

l

m

n

o

a

b

c

d

e

f

g

h

i

j

k

l

m

n

14.9 Larger transistor portables made in the United States with many U.S.-made parts, 1960–65. *a*, RCA 1T4H, 1960 (6-6/8″ high); *b*, General Electric P790A, 1960 (5-5/8″ wide); *c*, General Electric P815A, 1960 (6″ wide); *d*, RCA 4RG51, 1963 (6-1/2″ high); *e*, Arvin 66R58, 1965 (6″ high); *f*, Motorola XT18S, 1960 (10-1/4″ wide); *g*, Zenith Royal 710, 1960 (5-3/8″ high); *h*, Sylvania 2800, 1960 (6-3/4″ high); *i*, Zenith Royal 755, 1960 (9″ wide); *j*, Sylvania 4P19W, ca. 1961 (6-1/4″ wide); *k*, Zenith Royal 490, 1963 (6-3/4″ wide); *l*, Silvertone 600 (Sears 2215), 1961 (8-1/4″ wide); *m*, Zenith Royal 275, 1960 (5-3/4″ high); *n*, Philco T76-124, 1960 (7-1/8″ wide); *o*, Westinghouse H772P6GP, 1961 (7-1/2″ wide); *p*, Motorola CX2B, 1963 (5-5/8″ high); *q*, Silvertone 800 (Sears 2222–2223), 1962 (6-1/4″ high); *r*, General Electric P810A, 1962 (8″ wide); *s*, Silvertone 500 (Sears 2212), 1962 (7″ wide); *t*, General Electric P856A, ca. 1964 (8-1/2″ wide); *u*, Zenith Royal 790, ca. 1963 (9″ wide); *v*, Admiral Y2098, 1961 (5-7/8″ wide); *w*, RCA 3RG14, 1962 (6-3/8″ high); *x*, Motorola X17B, 1960 (5-3/4″ high); *y*, Silvertone 800 (Sears 4223), 1963 (10-1/2″ wide)

In the late fifties and very early sixties, U.S. companies continued to make larger transistor portables—especially lunch boxes—largely at home (Fig. 14.9). These were profitable items that initially were not under heavy import pressure. The assembly of transistor lunch boxes, even small ones, was unproblematic. Overall designs and manufacturing operations were identical to those perfected on tube portables. In fact, some sets had a metal chassis rather than a printed circuit board. Servicemen obviously rejoiced over this design, but such sets were not preparing U.S. manufacturers for the coming age of microelectronics. The mass production of transistor lunch boxes presented no technological challenges, and so stimulated no technological innovations. Though sounding good, American-made transistor lunch boxes with their large, American-made components seem crude inside compared to their Japanese counterparts (Fig. 13.4h). As import pressure increased through the sixties, many U.S.–brand lunch boxes were manufactured abroad or had mostly foreign-made parts.

Cheap labor was obviously the critical factor in Japan's total conquest of shirt-pocket radios, but creative marketing strategies also played a significant role. In the late fifties, Japanese companies distributed their radios through unconventional channels for electronic products, relying, for example, on distributors of novelty items, which had previously handled Japanese trinkets. As a result, these radios were available in drug stores, jewelry stores, department stores, discount stores, even hardware stores, where they were very visible to the ultimate consumer—teenagers.

The Japanese also sold sets to mass-merchandisers Sears, J. C. Penney, and Montgomery Ward, which marketed the radios under their own house brands. This was another big foot in the door. That Japanese companies offered shirt-pocket radios in more than seventy-five brands and more than two hundred models also boosted their market share. Adults, who bought many of the sets as gifts for their teenagers, in the late fifties had strong loyalties to American-made products. However, few people probably suspected that radios named "Americana," "Trans-America," and "Hudson" were made in Japan (Fig. 14.10a–c). To avoid the stigma of "Jap crap," a great many Japanese companies had adopted U.S.–sounding brand names. (Even Japanese transistor lunch boxes sometimes had misleading names, like Playmate and GM—Fig. 14.10d–e.) Other made-in-Japan radios wore respected U.S. brands like Motorola and Magnavox. Moreover, in magazine advertising, Japanese firms seldom identified themselves as such. Even if the Japanese origins of these radios had been obvious, the vast number of sets available in so many stores was a decisive advantage, for numerous people would have bought the first—or cheapest—set they encountered. Teenagers, of course, had little loyalty to American brands.

In the final analysis, many executives in U.S. electronics companies did not view loss of the shirt-pocket portable to the Japanese as a calamity or as an omen of worse things to come. After all, it was a low-profit child's radio. Thanks to military orders, the U.S. electronics industry had record sales and profits; the Japanese were taking a very small nibble out of a very large and rapidly growing pie. U.S. companies in the

d

e

14.10 Some Japanese transistor sets with brands that suggest an American origin. *a,* Trans-American SR-6T60, ca. 1961; *b,* Americana FP80, 1962; *c,* Americana FP64, 1961; *d,* Playmate, ca. 1963 (8″ wide); *e,* GM Sportsman, ca. 1963 (8″ wide)

early sixties (gearing up for a big color TV push) were confident that the Japanese, who were still heavily dependent on American technology, would not quickly master TV. And when they did, it was thought, they would concentrate for awhile on selling TVs at home in Japan, since so few people had them. Also, because the silicon integrated circuit had just been invented, which promised to streamline product assembly, U.S. firms believed they could maintain their lead in technology, thereby compensating for lower foreign wages.

All of this was wishful thinking, as a few thoughtful observers noted at the time. SONY shipped its first black-and-white TV, a transistorized portable, to the United States in 1960, the same year Japan began to produce color TVs. As in the case of the shirt-pocket radio, other Japanese companies quickly joined SONY in the U.S. marketplace, and consumers responded positively. Applying the lesson of the shirt-pocket portable—that head-on competition was in the end a losing proposition—American firms soon ceded manufacture of black-and-white sets to the Japanese. Color TVs followed in the 1970s (though many are now made by Japanese and Korean companies in the United States).

Portable radios of the Japanese period, regardless of maker or country of origin, display a marked functional diversity. To find these new niches in the market, radio makers experimented with new models.

In 1959 and 1960 several manufacturers attempted to overcome the limitations of the shirt-pocket set's small speaker and baffle by offering auxiliary speakers. When the radio was stationary (and not in the pocket), it could be plugged into the speaker, creating a sort of battery-powered table model (Fig. 14.11a). *Consumer Reports* tested these products and came away—as they often did—unimpressed. Their conclusion was blunt: "If you want a portable radio receiver that is more than a novelty, that can produce loud, clear, and satisfying music, you will have to get a larger, probably heavier, and perhaps more expensive battery-operated set." It is curious that the shirt-pocket portable was still regarded in 1960 by adults as a "novelty."

For reasons that do not seem obvious now, both U.S. and Japanese companies thought that consumers would go for battery-operated table models. A handful of "cordless table models" were manufactured during the late fifties and early sixties; they were about the size of traditional tube table models. Like portables, these sets were completely self-contained but, lacking handles, were intended to be sedentary (Fig. 14.11b–d). The most unusual model was Philco's T1000 of 1959 (Fig. 14.11b), obviously styled to match its Predicta televisions. The cordless table model genre did not catch on; having to change batteries in a stationary radio made little sense to most people. When transistor table models finally began their rise to popularity in the mid-sixties, they all had plugs (and were AC-only).

In another attempt to make the transistor portable more versatile, a few set makers developed portable car radios (reviving a genre that had died quietly in the thirties). These convertibles (Fig. 14.11e) enjoyed only modest success; after all, by the late fifties most cars came new with quite nice transistor radios.

a

b

c

d

e

14.11   An assortment of versatile transistor portables. *a,* Hitachi TH666R (radio) and ES904 (speaker), 1959 (6-7/8″ wide); *b,* Philco T-1000-24, 1959 (15-1/2″ wide); *c,* Admiral 561, 1959 (11″ wide); *d,* Channel Master 6511, 1962 (12-3/8″ wide); *e,* General Electric P871A, 1962 (7-5/16″ wide)

a

b

e

c

d

14.12   Some U.S.-made AM-FM transistor portables, 1961–65. *a,* Zenith Royal 2000, 1961 (8-3/8″ high); *b,* Arvin 62R98, 1962 (6-3/4″ high); *c,* Zenith Royal 820, 1964 (6-3/8″ high); *d,* Arvin 64R78, ca. 1964 (5″ high); *e,* General Electric P1818B, 1965 (6-7/8″ wide)

The first transistor AM-FM portables, of Japanese manufacture, began appearing in 1959. With ample reasons to expect that tiny sets would be snapped up by Americans, Japanese companies offered AM-FM radios in the micro-table-model style. However, these AM-FM portables, which sounded just awful, appealed to neither adults nor children. Of course larger AM-FM lunch boxes did find favor among consumers (Fig. 14.12).

In 1960, Zenith brought out the first AM-FM portable made in the United States (Fig. 14.12a). It was a huge and heavy lunch box, with a large speaker capable of making mellow music. The twelve-pound Zenith Royal 2000, with metal chassis and plug-in transistors, cost a whopping $150. A few years later, Zenith downsized and downpriced its AM-FM offerings. For example, the Royal 820 (Fig. 14.12c) was sold in 1965 for $80. Remarkably, it still had a metal chassis and was hand-wired.

When FM stations in the middle sixties began playing album rock, AM-FM radios enjoyed a surge in popularity. A large market opened up for AM-FM portables, and eventually stereo portables. The young adult listeners to these stations demanded good fidelity—rock's first generation had grown up—and so in the seventies was born the modern boom box. (In the eighties, that same generation would become the Walkman's first important market.)

In 1960 the Zenith Trans-Oceanic underwent another transformation, getting one more band and becoming the Royal 1000D (for $275). Three years later, FM was added to create the $200 Royal 3000 (Fig. 14.13). In the first days of August 1964 Zenith manufactured the one-millionth Trans-Oceanic, though McDonald did not live to see it. The final Trans-Oceanic model, the Royal 7000, was released in 1973. In 1981 the Trans-Oceanic (then being manufactured in Taiwan) was discontinued after four decades of distinguished service.

14.13   Zenith Trans-Oceanic Royal 3000, 1963 (12-1/2″ wide)

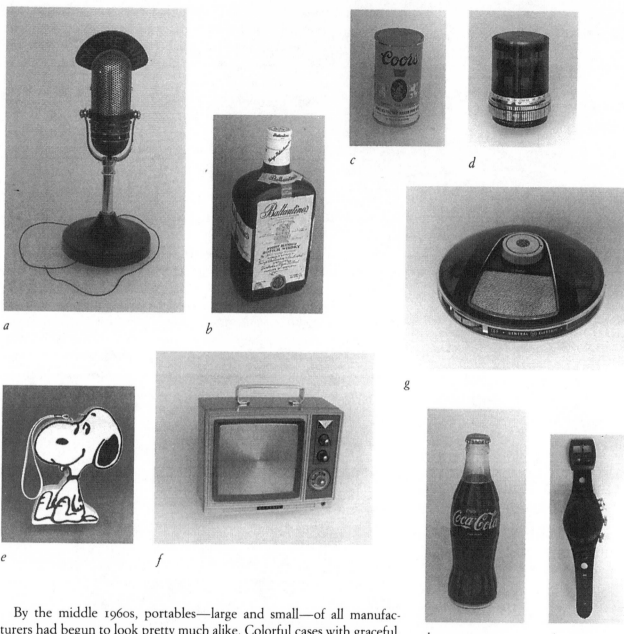

a   b   c   d   e   f   g   h   i

By the middle 1960s, portables—large and small—of all manufacturers had begun to look pretty much alike. Colorful cases with graceful, flowing lines were abandoned. In their place came the gray or black or tan plastic box, sometimes with chrome-colored trim that quickly wore off. In the eyes of their designers these sets were intended to evoke images of modernity, and perhaps they did.

There are some notable exceptions to the trend of blandness and sameness that came to characterize portable radios in the sixties. Because a transistor radio's guts were tiny, they could be placed into cases of odd sizes and shapes. As a result, designers created an endless variety of novelty sets: transistor radios that did not look like radios (Fig. 14.14). Though the novelty genre was not a novelty, the transistor radio gave it enormous possibilities, which were realized throughout the sixties, seventies, and eighties. Radios appeared in the guise of model cars, stuffed animals, flying saucers—and wristwatches. For many children, a novelty set—as a toy—became their first radio, and the crystal set faded into obscurity. Novelty sets also gave the portable radio an entirely new purpose in American life: it could be used for advertising. Sometimes a company like Havoline Oil or Ballantine's Scotch would order an entire run of novelty radios that represented one of their products. In a few cases, portables used for advertising were ordinary in all respects except for a conspicuous company name and logo. Usually, the radios inside novelty sets large and small were of the same quality as a cheap shirt-pocket portable and produced abysmal sound. However, as toys and mobile advertising, these radios did not need to be high fidelity.

In 1958, the first full year in which Japanese transistor radios were marketed in the United States, sales of all portables passed the five million mark. In 1960, nearly ten million were sold. Sales of portable radios in 1965 exceeded twenty-one million. At the end of the decade, in 1969, consumers bought twenty-seven million sets, and more than half were AM-FM.

Clearly, during the sixties, the portable radio found a place in nearly all American homes. As the price of portables continued to fall—a shirt-pocket set could be bought for one or two hours wages—people who previously knew nothing of portables began to buy them and soon found them indispensable. Portables were no longer confined to the trend-setters and immediate trend followers, or to young adults and teenagers. Instead, Americans of all ages, in all walks of life, and in all communities could be found enjoying various activities with their portable radio companion, much as Zenith had foreseen decades earlier. During those four decades, electronic technology and American life had undergone many and far-reaching changes, and portable radios had responded. Both Sarnoff and Gernsback lived to witness the triumph of the portable radio or, rather, its trivialization. For, by the end of the sixties, the portable radio—once a powerful symbol of technological prowess—had become ordinary.

Robust sales of portable radios was the good news; the bad news was that more and more of this market was being captured by imports. In 1960, already 54 percent of the portable radios sold in the United States were made abroad. By 1965, the figure was 67 percent, and in the year of Woodstock and the first moon walk, 1969, fully 94 percent of the portables sold were imports.

As production costs rose in Japan during the sixties, Japanese companies moved much of their radio making to other countries in the Pacific rim, where labor was still cheap. Ironically, little more than a decade after the Regency TR-1 and the SONY TR-63 appeared, not many small portable radios were being manufactured in the United States or Japan.

14.14   An assortment of novelty transistor radios, made in Japan, Hong Kong, and Taiwan (ca. late fifties to the eighties). *a,* Leadworks "On The Air" classic radio microphone (12-5/8″ high); *b,* Ballantine's Scotch whiskey bottle, ca. late fifties (8-3/4″ high); *c,* Coors beer can (4-3/4″ high); *d,* General Electric P2760, 1970 (4-1/4″ high); *e,* Snoopy (6-1/2″ high); *f,* Classic TV (9-1/2″ wide); *g,* General Electric P2775 (5-1/4″ wide); *h,* Coca-Cola bottle (7-7/8″ high); *i,* Amico wrist radio, ca. mid-sixties (2-3/8″ wide)

15

The Triumph of Cryptohistory

ONE DAY IN THE FALL OF 1980, ALTHOUGH NO ONE RE-
members exactly when, Zenith closed down an assembly line in Plant 2,
on Kostner Street in Chicago. Zenith had moved production of the
Trans-Oceanic Royal 7000 to its factory in Taiwan. Unnoticed at the time,
this event silently ended an era. Portable radios were no longer being
manufactured in the United States.

Today's portable radios come from many countries, mostly on the Pacific rim, including Korea, Taiwan, China, Singapore, and Malaysia. A
few are even made in Japan. Many brands are relative newcomers that
entered after the sixties, but one can still find some old friends, like SONY
and Toshiba.

What happened to the old-line American companies? Some, like
Motorola, Zenith, and Automatic, discontinued making radios entirely.
Motorola remains an important producer of communications gear, including cellular phones, some of which are actually assembled in the
United States; Zenith still sells TVs and VCRs, and is the only American-
owned company that assembles any TVs at home; Automatic has been
reduced to selling electric bug-zappers. Hoffman also dropped out of
consumer electronics, finding military contracting much more lucrative.
Admiral, which absorbed the remnants of Belmont in 1955, was purchased by North American Rockwell, a major military contractor, and
has not been heard from since.

Other well-known American brands adorn radios and TVs made
abroad by foreign-owned, multinational corporations. The Philco, Sylvania, and Magnavox names are the property of Philips of the Nether-
lands. Most amazing of all, the consumer electronics divisions of RCA
and GE are owned by Thomson, a French firm. Sarnoff would not be
pleased.

Today, multinational corporations, based here and abroad, are ag-
gressively cultivating markets around the globe. Indeed, the world is
being knit into a single consumer society dominated by huge, faceless
corporations, from SONY to General Motors. Significantly, ads for consumer products seldom mention where the company is based, much less

where its products are actually made. This is not surprising. Traditional allegiances to countries stand in the way of global manufacturing and marketing strategies. What the new corporate colossi wish to foster is, above all, loyalty to themselves.

Multinational corporate strategies to break down traditional allegiances include the use of cryptohistory. Employing the objects of everyday life, multinationals appropriate to themselves credit for significant technological accomplishments, deliberately divorcing them from the individuals, companies, and countries that, in fact, brought them into being. Cryptohistory is creating a new past to go along with the new future that multinationals are ushering in. Such cryptohistory is disseminated through ads, to be sure, but more insidiously by unwitting—and sometimes unthinking—third parties, particularly journalists. Indeed, the mass media are helping to make the world safe for the multinationals. No product illustrates this process better than the portable radio, especially the shirt-pocket portable.

On January 8, 1989, the popular CBS program "60 Minutes" did a piece entitled "Mr. SONY." Diane Sawyer interviewed Akio Morita on the ills of American business. In introducing Morita, Sawyer said that "SONY can . . . take credit for the first pocket radio." In a later voice-over, Sawyer repeated the blunder: "[SONY] took an American invention, the transistor, and added a stroke of practical genius to produce the pocket radio." Morita could have corrected Sawyer (some SONY publications do acknowledge the priority of the Regency TR-1), but there is no evidence that he did.

How could a respected television journalist make such a mistake? Perhaps Sawyer relied on Nick Lyons' book, *The Sony Vision,* published in 1976. In this gushy, wide-eyed history of SONY, Lyons claimed that the SONY TR-63, of 1957, was "the world's first pocket-size all-transistor radio." Another book that could have led Sawyer astray is John Heskett's *Industrial Design,* in which the SONY TR-63 was also said to be "the first 'shirt-pocket' radio." Similarly, in *CEO,* the authors recount the story of how IBM's Thomas J. Watson, Jr., used transistor radios to motivate the company's computer engineers. In their telling the Regency TR-1 became a *Japanese* transistor radio. Obviously, there is no lack of portable radio cryptohistory that could have influenced Sawyer.

When I talked to spokespersons for "60 Minutes," they insisted—rather defensively—that no errors had been made. I was not the first to correct them, however. One Mrs. R. L. Campbell took out a small display ad in the *Wall Street Journal.* Obviously exercised, she stated that the transistor pocket radio was actually her late husband's handiwork. He demonstrated it, the ad continued, at an engineering convention in 1953. There is no reason to doubt her contention, for electronics experimenters were building transistor apparatus as soon as they could get their hands on the tiny amplifiers (Chapter 12). The CBS officials I talked to denied knowledge of Mrs. Campbell or her *Wall Street* ad. Exasperated, I wrote to D. Hewitt, executive producer of "60 Minutes," pointing out that Americans had made the first shirt-pocket portables, in both their tube and transistor incarnations, and requested that "60 Minutes" correct their mistake. Mr. Hewitt did not reply. The ethics and journalistic standards of CBS news apparently have slipped somewhat from the halcyon days of Edward R. Murrow.

CBS is not the only network that gives SONY free commercials and misrepresents the portable radio's history. On August 18, 1989, NBC's *Today* show did a lengthy piece on the tenth anniversary of SONY's Walkman. Akio Morita told the story of how he came up with the idea for this product, which differs from the version in his book, *Made in Japan*. In the newer story, he recalled noting how the boom boxes common in the late seventies were just too bulky to be used conveniently. The Walkman, then, was an attempt to make radio-cassette players more portable. To younger viewers, unfamiliar with earlier shirt-pocket radios and pocket cassette recorders, the story strongly implied that SONY alone invented small portable radios and cassette players.

The BBC in collaboration with the U.S. Arts and Entertainment Network produced a history of Japan entitled "Nippon." Shown in the United States in January 1991, it included an interview with SONY's other living ancestor, Masaru Ibuka. After describing how the transistor was invented by AT&T and licensed by SONY, the narrator reported that "before any American manufacturers, Dr. Ibuka had adapted the device and managed to make transistor radios." Examining a SONY TR-63, Ibuka himself added, "This is the first pocket transistor radio."

Even programs shown on educational television wallow in cryptohistory. In the series *Japan,* made for Chicago TV station WTTV, Narrator Jane Seymour in 1989 attributed to the Japanese "the world's first transistor radio." This product, the program stressed, resulted from the traditional concern in Japan with miniaturization.

In one medium after another, the portable radio's real history has been shaped to suit the purposes of SONY and other (mostly Japanese) multinationals. This use of cryptohistory, of course, is not new. We have seen that Philco and other U.S. companies in 1939 took credit for inventing the portable radio. More recently, I.D.E.A. ads claimed that the Regency TR-1 was the first pocket radio. We have become accustomed to corporations jockeying for market share with their refurbished pasts. It might be argued that if cryptohistory merely helps a company compete, then little harm has been done. After all, beyond antique radio collectors, who cares about the real history of the portable radio? Does it matter? I submit that, in fact, it matters a lot. Because ignorance of real product history is so widespread, cryptohistory has a way of insinuating itself into serious discussions of industrial and economic policy. Indeed, cryptohistory of the portable radio now contaminates debates about the causes of—and solutions to—America's apparent industrial decline. Far-reaching policy recommendations being considered by the U.S. government rest on a foundation of seemingly innocuous little white lies. Let us see how this came about.

It has taken a long time for people outside the industries shrunk by foreign competition to appreciate that America's manufacturing base has been eroding at a frightening rate. The U.S. government itself was unconcerned until the late seventies and eighties, when important national statistics—balance of payments, federal deficit, and so forth—turned uniformly unfavorable. Washington did finally take action of a sort: numerous studies were commissioned, which led to numerous weighty tomes. In addition, a small cottage industry of academics arose (mostly economists and political scientists), who also scrutinized the loss of American industry, often in mind-boggling detail. The consumer electronics industry, because it was the first and most thoroughly affected by foreign competition, has been the most important case, the focus of many books and reports now being pondered by public officials.

These studies make fascinating as well as depressing reading to people familiar with the early history of U.S.–Japanese competition in consumer electronics. This competition, of course, was over the shirt-pocket portable (Chapter 14). Somehow, even distinguished economists and political scientists have been taken in by portable radio cryptohistory. Over and over again they erroneously claimed that the U.S. industry did not fully appreciate the transistor's potential for making portable radios (and other consumer products) and failed to respond to the challenge of Japanese transistor portables, quickly surrendering the market without offering competitive products.

In 1981, in one of the earliest and most influential books on the problems faced by the U.S. consumer electronics industry, *The Japanese Challenge to U.S. Industry*, Jack Baranson wrote:

> In the early 1950s, U.S. component makers were moving within a limited production of transistor radios, built as electronic curiosities. The major U.S. radio manufacturers had traditionally built large console and table radios, had major investments in tube radio production facilities, and in most cases were also major producers of tubes. Thus they saw little incentive for pursuing the new technology extensively. The Japanese government and its fledgling electronics industry, however, saw the small transistor radio as a unique opportunity to enter the world market and earn badly needed foreign exchange. . . . A key ingredient in these early Japanese successes was the lack of competition from the major U.S. consumer electronic companies.

These misstatements about the past help to justify proposals for restructuring the economic environment in which U.S. companies operate.

According to Baranson and others, American firms did not challenge the Japanese imports with competitive products because the U.S. tax structure provided inadequate incentives for investing in research and product development. Thus, Japanese companies quickly took the lead in adopting labor-saving technologies for making transistor products. (In fact, just the reverse took place.) Obviously, the U.S. tax code must be changed to provide more write-offs for manufacturing companies.

It is also frequently observed that, in contrast to the U.S. situation, Japanese government and industry cooperate closely. The solution, proposed most recently by Susan Walsh Sanderson in *The Consumer Electronics Industry and the Future of American Manufacturing: How the United States Lost the Lead and Why We Must Get Back in the Game,* is to modify American anti-trust law (enacted originally to combat the robber barons). A relaxation of anti-trust laws will, it is thought, promote collaboration among companies and create harmonious relationships be-

tween government and industry. In a more favorable economic environment, U.S. companies should be able to catch up to Japan in the technology of product assembly.

It is obvious that tampering with the tax code and anti-trust laws can
create conditions more congenial to U.S. corporations, especially multinational corporations (wherever they are based). But will a welfare program for multinationals actually induce American firms to resume (or
take up) the manufacture of consumer electronic products in the United
States? My answer, based on the lessons of real product history, regrettably is no.

Contemporary analysts assume that assembly of consumer electronic
products is now exclusively a high-technology, capital-intensive enterprise. Since manufacturing operations would be highly automated, cheap
labor should not furnish a comparative advantage to foreign firms. With
enough robots U.S. companies really can compete. The history of
American-Japanese competition in shirt-pocket portables suggests otherwise. Motorola, the industry leader in using labor-saving technology in
the fifties, was one of the first companies to transfer production of shirt-
pocket sets offshore (to take advantage of low-priced labor in Japan). In
fact, product assembly in consumer electronics can be labor-intensive,
technology-intensive, or something in between. Cheap labor obviously
can substitute for high technology; Motorola saw that very quickly, and
other U.S. radio makers soon learned this lesson the hard way. It is a
lesson fully understood today by Japan's and America's ablest global
competitors, which rely on a mix of high-tech and high-labor operations,
sometimes spread among factories in different countries.

That cheap foreign labor may be easily substituted for automated assembly has clear-cut implications for U.S. firms desiring to be competitive with new products. If an American company introduces a wondrous
new gadget, built on a high-tech assembly line, multinationals will
reverse-engineer it, farming out certain manufacturing operations to sub-
sidiaries in low-wage countries. The foreign copy (perhaps put together
in Singapore or China using components made in Japan, Taiwan, and
Korea) will enter the U.S. market priced below the original. Despite tax
incentives for research, development, and automation, and regardless of
government-business harmony, the U.S.–made product will soon be at a
competitive disadvantage at home and abroad. Even if the U.S. components industry were to revive, overall cost differentials would still favor
the multinationals' clone.

Let us suppose, however, that the American entrepreneur has been
very clever and squeezed out of the product every iota of labor. The
factory would consist of a small army of robots making the parts, which
would be fed (mechanically, of course) to another platoon of robots that
assembles and packages the final product. (How this solution would put
Americans back to work in manufacturing is another issue.) Ostensibly,
foreign multinationals would be unable to compete, pricewise (assuming
large enough production runs to justify the high initial outlays for the
robots). Again, lessons learned from real product history suggest caution
in coming to optimistic conclusions. Beginning with SONY's earliest entry
into the U.S. market, Japanese firms in particular have shown a willing-

ness to suffer losses, sometimes over long periods, in order to maintain a presence and to garner, even if slowly, a significant market share. U.S. companies learned in the early sixties that a lengthy price war would not be to their advantage; Japanese firms still had profits at home to underwrite their price wars abroad. There is no question that, in a prolonged price war today, Toshiba or Matsushita would outlast an American rival, particularly if the latter were a small company.

Thus, the remedies proposed to reverse the decline of the U.S. consumer electronics industry—resting on a foundation of cryptohistory—cannot possibly work. The lessons of real history, which U.S. firms appreciated by the early sixties, are unequivocal: the American market for any new U.S.–made product will be invaded—perhaps eventually captured—by multinationals selling cheaper, foreign-made clones.

This assessment is bleak, but there is hope. The key to rebuilding the consumer electronics industry at home is the judicious use of tariffs to protect *new* products. Sanderson, Baranson, and most other contemporary analysts write off tariffs as being at odds with U.S. free-trade policy and therefore counterproductive. Though clearly anathema to multinationals and to free-trade doctrinaires, tariffs can work. After all, Japan and most other industrial nations use them rather effectively.

Instead of protecting aging and inefficient industries and mature products, tariffs can be employed to insure that U.S. entrepreneurs who come up with innovative products will be able to reap the rewards in a protected home market. During that time (say fifteen years), tariffs on clones, copies, and other obvious derivatives of that product would be set high enough to exclude the imports completely. Such tariffs would protect only new products made by U.S.–owned companies (not multinationals) that do their manufacturing in the United States, substantially with U.S.–made parts, substantially using U.S.–made equipment. Each new product, then, can generate markets for U.S. suppliers. Most importantly, a large and vigorous entrepreneurial fringe can arise because financiers will see profit potential in small companies that have good product ideas.

Could the United States become competitive in the global marketplace? I believe so. Multinationals—foreign and domestic—would be motivated to collaborate rapidly and fairly with the U.S. firm to license its new product for sales abroad (and perhaps foreign manufacture). Thus, multinationals will still be heavily involved in disseminating new products in the *global* marketplace. They will also be eager to establish close ties with American-owned entrepreneurial firms, perhaps underwriting research and development costs of promising products in exchange for foreign marketing rights.

Another recommendation is that for any product sold in the United States the country (or countries) of manufacture should be prominently identified—on the item itself and in *all* advertising.

In one important respect I do agree with mainstream analysts of U.S. industrial decline: American economic policy does not effectively promote the growth of a healthy consumer electronics industry. My recommendations, however, do not give privileged status to large multinationals; rather, they favor smaller, U.S.–owned firms that develop and manufacture new products at home, thereby creating jobs *at home* and markets for other U.S. manufacturers. Economic policy should now begin to tilt toward companies that expand instead of export employment.

The shirt-pocket portable was a cultural imperative throughout the teens, twenties, and thirties, a spectacular failure in the forties, and a runaway success beginning in the late fifties. Today it has also become a symbol with different meanings for different groups. To Mrs. Campbell and the creative tinkerers she represents, the shirt-pocket portable stands for the American genius, ahead of his time, who receives inadequate recognition for his achievements.

For many Japanese, the shirt-pocket portable has come to mark the beginning of a new industrial era in which their electronics companies could compete on an equal footing with U.S. companies in the U.S. marketplace. This was a remarkable turning point, a mere dozen years beyond the terror of Hiroshima and Nagasaki. The shirt-pocket portable also represents the beginnings—however tentative—of consumer microelectronics, an industry the Japanese now thoroughly dominate. This tiny radio, then, scarcely fit to play music, has become to Japan's senior generation a powerful symbol of national pride.

This very same product has quite different meanings to Americans familiar with the real history of consumer electronics. It calls to mind the heroic efforts of Texas Instruments and I.D.E.A. to produce rapidly and market the world's first transistor radio. The achievement of the Regency TR-I exemplifies what American engineers can accomplish under high pressure through teamwork.

A second meaning to that group is somewhat more gloomy. The first American transistor radios spawned a host of cheap foreign imitations, which then invaded—and captured—the U.S. portable radio market. The loss of transistor radios was followed by black-and-white TV, color TV, and stereos. No American company makes its own VCRs, and when high-definition television arrives in the 1990s, American companies, which were experimenting with the technology beginning in the 1940s, will probably be watching on the sidelines. The shirt-pocket portable, then, represents the start of a long decline in America's consumer electronics industry.

In the late 1950s and early 1960s, U.S. policy makers paid no attention to the eclipse of the American shirt-pocket portable. After all, this type of set was mostly a toy, a novelty, through which teenagers played their infernal music. American adults had greater things to occupy their minds, and efforts, such as reaching the moon by 1970 and recapturing the supposed Soviet lead in science represented by the 1957 launch of Sputnik. Today, however, the shirt-pocket portable has become to policy makers a metaphor for the beginning of the end. Clearly, this small article carries a heavy load of meaning.

Though poor performers overall and without noteworthy circuitry, the shirt-pocket portables of the late 1950s and early 1960s are now collectibles. These sets have attractive plastic cases that call to mind their middle-aged owners' teenage years. For them, these radios represent a different era, filled with the hopes of youth and the aspirations of a generation, an idealism brutally crushed by the Vietnam War. The tiny radio also brings to mind those playful times in class and after bedtime when listening (through the ear plug) was just a little bit naughty.

Because we Americans do not know the origin of everyday things, it is easy for us to fall prey to propaganda, to accept as truth the myths that promote the interests of multinational corporations and other countries. The myths that U.S. companies did not see the transistor's potential for consumer products and failed to mount a credible challenge to Japanese imports (and that SONY originated the shirt-pocket radio) together serve a pragmatic purpose: they make the Japanese domination of consumer electronics seem to be part of the natural order—something that cannot be changed. Such beliefs deny the rich heritage of American invention (that thrives even to this day) in tinkerers' workshops, universities, government laboratories, and corporations—large and small. If we lose sight of the true history of the shirt-pocket portable today, perhaps tomorrow a generation of Americans will grow up believing that the Japanese invented television or the first portable TV, or even color TV. Such Americans are apt to accept passively and as inevitable the decline of other American consumer product industries. To surrender the past, then, is to surrender the future.

SELECTED BIBLIOGRAPHY

AITKEN, HUGH G. J.
  1976   *Syntony and spark—the origins of radio.* Princeton, New Jersey: Princeton University Press.
  1985   *The continuous wave: technology and American radio, 1900–1932.* Princeton, New Jersey: Princeton University Press.

ALLEN, FREDERICK LEWIS
  1952   *The big change: America transforms itself, 1900–1950.* New York: Harper and Brothers.

ALLEN, G. C.
  1958   *Japan's economic recovery.* London: Oxford University Press.

ANDREWS, ALLEN
  1977   *The flying machine: its evolution through the ages.* New York: G. P. Putnam's Sons.

ARMYTAGE, W.H.G.
  1968   *Yesterday's tomorrows: a historical survey of future societies.* Toronto: University of Toronto Press.

BARANSON, JACK
  1981   *The Japanese challenge to U.S. industry.* Lexington, Massachusetts: Lexington Books.

BARNOUW, ERIK
  1966   *A tower in Babel: a history of broadcasting in the United States, Volume I—to 1933.* New York: Oxford University Press.
  1968   *The golden web: a history of broadcasting in the United States, Volume II—1933 to 1953.* New York: Oxford University Press.
  1970   *The image empire: a history of broadcasting in the United States, Volume III—from 1953.* New York: Oxford University Press.

BASALLA, GEORGE
  1988   *The evolution of technology.* Cambridge: Cambridge University Press.

BELLAMY, EDWARD
  1887   *Looking backward, 2000–1887.* New York: Ticknor and Company.

BERGER, ARTHUR ASA
  1973   *The comic-stripped American.* New York: Walker and Company.

BLUESTONE, BARRY, AND BENNETT HARRISON
1982   *The deindustrialization of America: plant closings, community abandonment, and the dismantling of basic industry.* New York: Basic Books.

BOLLING, RICHARD, AND JOHN BOWLES
1982   *America's competitive edge: how to get our country moving again.* New York: McGraw-Hill.

BRADEN, DONNA R.
1988   *Leisure and entertainment in America.* Dearborn, Michigan: Henry Ford Museum and Green Field Village.

BRAUN, ERNEST, AND STUART MACDONALD
1978   *Revolution in miniature: the history and impact of semiconductor electronics.* Cambridge, Massachusetts: Cambridge University Press.

BUCKLEY, ROGER
1985   *Japan today.* Cambridge, Massachusetts: Cambridge University Press.

CALDER, NIGEL (EDITOR)
1965   *The world in 1984: volumes 1 and 2. The complete new scientist series.* Harmondsworth, England: Penguin Books.

CARTER, PAUL A.
1977   *The creation of tomorrow: fifty years of magazine science fiction.* New York: Columbia University Press.

CHENEY, MARGARET
1981   *Tesla: man out of time.* Englewood Cliffs, New Jersey: Prentice-Hall.

CLARK, GEORGE H.
1946   *The life of John Stone Stone.* San Diego, California: Frye and Smith.

COHEN, STEPHEN S., AND JOHN ZYSMAN
1987   *Manufacturing matters: the myth of the post-industrial economy.* New York: Basic Books.

CORN, JOSEPH J. (EDITOR)
1986   *Imagining tomorrow: history, technology, and the American future.* Cambridge, Massachusetts: The MIT Press.

CORN, JOSEPH J., AND BRIAN HORRIGAN
1984   *Yesterday's tomorrows: past visions of the American future.* New York: Summit Books.

CUTCLIFFE, STEPHEN H., AND ROBERT C. POST
1989   *In context: history and the history of technology. Essays in honor of Melvin Kranzberg.* Bethlehem, Pennsylvania: Lehigh University Press.

DAVENPORT, BASIL, ROBERT A. HEINLEIN, C. M. KORNBLUTH, ALFRED BESTER, AND ROBERT BLOCH
1964   *The science fiction novel: imagination and social criticism.* Chicago: Advent Publishers.

DAVIDSON, WILLIAM H.
1984   *The amazing race: winning the technorivalry with Japan.* New York: John Wiley and Sons.

DAVIS, HENRY B.O.
 1983 *Electrical and electronic technologies: a chronology of events and inventors from 1900 to 1940.* Metuchen, New Jersey: Scarecrow Press.

DE FOREST, LEE
 1950 *Father of radio: the autobiography of Lee de Forest.* Chicago: Wilcox and Follett.

DE MENTE, BOYE, AND FRED T. PERRY
 1968 *The Japanese as consumers: Asia's first great mass market.* New York: John Weatherhill.

DERTOUZOS, MICHAEL L., RICHARD K. LESTER, ROBERT M. SOLOW, AND THE MIT COMMISSION ON INDUSTRIAL PRODUCTIVITY
 1989 *Made in America: regaining the productive edge.* Cambridge, Massachusetts: MIT Press.

DOUGLAS, ALAN
 1988 *Radio manufacturers of the 1920's, vol. 1—A-C Dayton to J. B. Ferguson, Inc.* Vestal, New York: The Vestal Press.
 1989 *Radio manufacturers of the 1920's, vol. 2–Freed-Eisemann to Priess.* Vestal, New York: The Vestal Press.

DOUGLAS, GEORGE H.
 1987 *The early days of radio broadcasting.* Jefferson, North Carolina: McFarland and Company.

DOUGLAS, SUSAN J.
 1987 *Inventing American broadcasting, 1899–1922.* Baltimore: The Johns Hopkins University Press.

DREHER, CARL
 1977 *Sarnoff: an American success.* New York: Quadrangle/The New York Times Book Co.

DU VALL, NELL
 1988 *Domestic technology: a chronology of developments.* Boston, Massachusetts: G. K. Hall and Co.

EISEN, JONATHAN (EDITOR)
 1969 *The age of rock: sounds of the American Cultural Revolution.* New York: Vintage Books.

ELECTRONIC INDUSTRIES ASSOCIATION
 1970 *Electronic market data book, 1970.* Washington, D. C.: Marketing Services Department, Electronics Industries Association.

ENTWISTLE, BASIL
 1985 *Japan's decisive decade: how a determined minority changed the nation's course in the 1950s.* London: Grosvenor Books.

EVERITT, C.W.F.
 1975 *James Clerk Maxwell: physicist and natural philosopher.* New York: Charles Scribner's Sons.

FESSENDEN, HELEN M.
 1940 *Fessenden: builder of tomorrows.* New York: Coward-McCann.

FOX, RICHARD WIGHTMAN, AND T.J. JACKSON LEARS
 1983 *The culture of consumption: critical essays in American history, 1880–1980.* New York: Pantheon Books.

FRANKLIN, H. BRUCE

1966 *Future perfect: American science fiction of the nineteenth century.* New York: Oxford University Press.

FRITH, SIMON

1981 *Sound effects: youth, leisure, and the politics of rock 'n' roll.* New York: Pantheon Books.

FULLER, ALVARADO M.

1890 *A.D. 2000.* Chicago: Laird and Lee.

GIBBS-SMITH, CHARLES H.

1970 *Aviation: an historical survey from its origins to the end of World War II.* London: Her Majesty's Stationery Office.

GILBERT, HORACE D. (EDITOR)

1961 *Miniaturization.* New York: Reinhold Publishing Corp.

GOLDMAN, MARTIN

1983 *The demon in the aether: the story of James Clerk Maxwell.* Edinburgh: Paul Harris Publishing.

GOLDMARK, PETER C., AND LEE EDSON

1973 *Maverick inventor: my turbulent years at CBS.* New York: Saturday Review Press/E.P. Dutton.

GOULD, CHESTER, AND HERB GALEWITZ

1978 *Dick Tracy, the thirties: tommy guns and hard times.* New York: Chelsea House.

GOULD, RICHARD A., AND MICHAEL B. SCHIFFER (EDITORS)

1981 *The Archaeology of U.S.* New York: Academic Press.

GRAHAM, LAWRENCE, AND LAWRENCE HAMDAN

1987 *Youth Trends: capturing the $200 billion youth market.* New York: St. Martin's.

GREGORY, GENE

1986 *Japanese electronics technology: enterprise and innovation.* Chichester: John Wiley and Sons.

GRINDER, ROBERT E., AND GEORGE H. FATHAUER

1986 *The radio collector's directory and price guide.* Scottsdale, Arizona: Ironwood Press.

GROSSMAN, LOYD

1976 *A social history of rock music from the greasers to glitter rock.* New York: David McKay Company.

GUILLAIN, ROBERT

1970 *The Japanese challenge.* New York: J. B. Lippincott.

HALDANE, J.B.S.

1924 *Daedalus or, science and the future.* New York: E. P. Dutton.

HAMMONTREE, PATSY GUY

1985 *Elvis Presley, a bio-bibliography.* Westport, Connecticut: Greenwood Press.

HESKETT, JOHN

1980 *Industrial design.* London: Thames and Hudson.

HOLLAND, MAX

1989 *When the machine stopped: a cautionary tale from industrial America.* Boston: Harvard Business School Press.

HOROWITZ, DANIEL

1985 *The morality of spending: attitudes toward the Consumer Society in*

America, 1875–1940. Baltimore: The Johns Hopkins University Press.

HUGHES, THOMAS P. (EDITOR)
  1975 *Changing attitudes toward American technology*. New York: Harper and Row.

HUGHES, THOMAS P.
  1989 *American genesis: a century of invention and technological enthusiasm, 1870–1970*. New York: Viking.

INGE, M. THOMAS (EDITOR)
  1982 *Concise histories of American popular culture*. Westport, Connecticut: Greenwood Press.

JOHNSON, CHALMERS
  1982 *MITI and the Japanese miracle: the growth of industrial policy, 1925–1975*. Stanford, California: Stanford University Press.

KANDO, THOMAS M.
  1975 *Leisure and popular culture in transition*. St. Louis: C. V. Mosby.

KAPLAN, MAX
  1960 *Leisure in America: a social inquiry*. New York: John Wiley and Sons.

KIKUCHI, MAKOTO
  1983 *Japanese electronics: a worm's-eye view of its evolution*. Tokyo: Simul Press.

KOTLER, PHILIP, LIAM FAHEY, AND SOMKID JATUSRIPITAK
  1985 *The New competition*. Englewood Cliffs, New Jersey: Prentice-Hall.

LARRABEE, ERIC, AND ROLF MEYERSOHN (EDITORS)
  1958 *Mass leisure*. Glencoe, Illinois: The Free Press.

LESSING, LAWRENCE
  1969 *Man of high fidelity: Edwin Howard Armstrong*. New York: Bantam Books.

LUNDWALL, SAM J.
  1971 *Science fiction: what it's all about*. New York: Ace Books.

LYONS, NICK
  1976 *The Sony vision*. New York: Crown Publishers.

MCELVAINE, ROBERT S.
  1984 *The Great Depression: America, 1929–1941*. New York: Times Books.

MACLAURIN, W. RUPERT, AND R. JOYCE HARMAN
  1949 *Invention and innovation in the radio industry*. New York: MacMillan.

MCMAHON, MORGAN E.
  1973 *Vintage radio: a pictorial history of wireless and radio* (second edition). Palos Verdes Peninsula, California: Vintage Radio.
  1975 *A flick of the switch, 1930–1950*. Palos Verdes Peninsula, California: Vintage Radio.
  1981 *Radio collector's guide, 1921–1932*. Palos Verdes Peninsula, California: Vintage Radio.

MEIKLE, JEFFREY L.
  1979 *Twentieth Century Limited: industrial design in America 1925–1939*. Philadelphia: Temple University Press.

MELMAN, SEYMOUR

1965   *Our depleted society.* New York: Holt, Rinehart and Winston.

1974   *The permanent war economy: American capitalism in decline.* New York: Simon and Schuster.

MITI

1960   *Transistor radio industry in Japan.* Tokyo: Ministry of International Trade and Industry.

MORITA, AKIO, EDWIN M. REINGOLD, AND MITSUKO SHIMOMURA

1986   *Made in Japan: Akio Morita and SONY.* New York: E. P. Dutton.

MOSKOWITZ, SAM (EDITOR)

1971   *Ultimate world: Hugo Gernsback.* New York: Walker and Co.

NAYAK, P. RANGANATH, AND JOHN M. KETTERINGHAM

1986   *Breakthroughs!* New York: Rawson Associates.

OFFICE OF TECHNOLOGY ASSESSMENT

1983   *International competitiveness in electronics.* Washington, D. C.: U. S. Congress, Office of Technical Assessment.

OGBURN, WILLIAM F., JOHN MERRIAM, AND EDWARD C. ELLIOTT (EDITORS)

1937   *Technological trends and national policy: including the social implications of new inventions.* Washington, D. C.: U. S. Government Printing Office.

O'NEILL, JOHN J.

1981   *Prodigal genius: the life of Nikola Tesla.* Hollywood, California: Angriff Press.

OZAWA, TERUTOMO

1974   *Japan's technological challenge to the West, 1950–1974: motivation and accomplishment.* Cambridge: MIT Press.

PAPER, LEWIS J.

1987   *Empire: William S. Paley and the making of CBS.* New York: St. Martin's Press.

PERRIS, ARNOLD

1985   *Music as propaganda: art to persuade, art to control.* Westport, Connecticut: Greenwood Press.

PULOS, ARTHUR J.

1983   *American design ethic: a history of industrial design to 1940.* Cambridge, Massachusetts: MIT Press.

RAE, JOHN B.

1984   *The American automobile industry.* Boston: Twayne Publishers.

RAMO, SIMON

1980   *America's technology slip.* New York: John Wiley and Sons.

RATHJE, WILLIAM L., AND MICHAEL B. SCHIFFER

1982   *Archaeology.* New York: Harcourt Brace Jovanovich.

REISCHAUER, EDWIN O., AND ALBERT M. CRAIG

1978   *Japan: tradition and transformation.* Boston: Houghton Mifflin.

ROEMER, KENNETH M.

1976   *The obsolete necessity: America in Utopian writings, 1888–1900.* Kent, Ohio: Kent State University Press.

SANDERSON, SUSAN WALSH

1989   *The consumer electronics industry and the future of American man-*

ufacturing: how the U.S. lost the lead and why we must get back in the game. Washington, D. C.: Economic Policy Institute.

SARNOFF, DAVID
1946   *Pioneering in television: prophecy and fulfillment.* New York: Radio Corporation of America.
1968   *Looking ahead: the papers of David Sarnoff.* New York: McGraw-Hill.

SCOTT, OTTO J.
1974   *The creative ordeal: the story of Raytheon.* New York: Atheneum.

SEGAL, HOWARD P.
1985   *Technological utopianism in American culture.* Chicago: University of Chicago Press.

SIEGEL, MARK
1988   *Hugo Gernsback: father of modern science fiction, with essays on Frank Herbert and Bram Stoker.* San Bernardino, California: The Borgo Press.

STOKES, JOHN W.
1982   *70 years of radio tubes and valves.* Vestal, New York: Vestal Press.

THOMAS, CHAUNCEY
1891   *The crystal button or, adventures of Paul Prognosis in the forty-ninth century.* Boston: Houghton, Mifflin and Company.

THOMPSON, SIR GEORGE
1955   *The foreseeable future.* Cambridge, Massachusetts: Cambridge University Press.

TOLSTOY, IVAN
1981   *James Clerk Maxwell: a biography.* Edinburgh: Canongate.

TSURUMI, YOSHI
1976   *The Japanese are coming: a multinational interaction of firms and politics.* Cambridge, Massachusetts: Ballinger.

TYNE, GERALD F.J.
1977   *Saga of the vacuum tube.* Indianapolis: Howard W. Sams and Company.

ULLMAN, JOHN E. (EDITOR)
1970   *Potential civilian markets for the military-electronics industry.* New York: Praeger Publishers.

VATTER, HAROLD G.
1963   *The U.S. economy in the 1950s: an economic history.* New York: W. W. Norton.

VOIGT, DAVID Q.
1974   *America's leisure revolution: essays in the sociology of leisure and sports.* (Revised edition). Reading, Pennsylvania: Albright College.

WHITE, WILLIAM J.
1976   *Airships for the future.* New York: Sterling Publishing Co.

WOLFF, MICHAEL F.
1985   The secret six-month project.   *IEEE Spectrum,* December.

YOSHINO, M. Y.
1971   *The Japanese marketing system: adaptations and innovations.* Cambridge, Massachusetts: MIT Press.

ILLUSTRATION CREDITS

Unattributed photographs are the author's.

1.1 Gernsback Publications, Inc.
2.1 *Radio News,* June 1924
2.2 Archives Center, National Museum of American History (NMAH), Smithsonian Institution, Neg. No. 87-9242
3.1 Archives Center, NMAH, Smithsonian Institution, Neg. No. 86-4085
3.2a Archives Center, NMAH, Smithsonian Institution, Neg. No. 80-14487
3.2b *Radio News,* December 1925
3.3 Archives Center, NMAH, Smithsonian Institution, Neg. No. 90-10154
3.4 Archives Center, NMAH, Smithsonian Institution, Neg. No. 90-10146
3.5 Archives Center, NMAH, Smithsonian Institution, Neg. No. 87-9255
3.6 *The Electrical Age,* June 1904
3.7 *Scientific American,* July 20, 1901
3.8 *Modern Electrics,* July 1909
3.9a Archives Center, NMAH, Smithsonian Institution, Neg. No. 90-10148
3.9b Archives Center, NMAH, Smithsonian Institution, Neg. No. 90-10147
3.10 Archives Center, NMAH, Smithsonian Institution, Neg. No. 90-10145
4.1 Archives Center, NMAH, Smithsonian Institution, Neg. No. 84-674
4.2a *Modern Electrics,* June 1909
4.2b *Modern Electrics,* April 1909
4.2c *Modern Electrics,* April 1909
4.3 *Scientific American,* January 13, 1906
4.4 *Scientific American,* June 8, 1895
4.5, *above Modern Electrics,* August 1909
4.5, *below Modern Electrics,* 1912
4.6a *Modern Electrics,* May 1912
4.6b *Modern Electrics,* May 1913
4.6c *Modern Electrics,* July 1913
4.6d *Modern Electrics,* January 1913
4.6e *The Electrical Experimenter,* late 1915 or early 1916
4.7 Archives Center, NMAH, Smithsonian Institution, Neg. No. 90-10143
4.8 *The Designer,* January 1918
5.1, *below Radio News,* December 1924
5.2 *Radio Merchandising,* June 1922
5.3 *Radio News,* December 1924
5.4a *Radio Broadcast,* July 1924
5.4b *Radio Age,* April 1924
5.5 Archives Center, NMAH, Smithsonian Institution, Neg. No. 90-10153
5.6 *Radio News,* June 1926
5.7 *Colliers,* November 20, 1926
5.8 *Radio News,* October 1925
5.9 *Radio Broadcast,* July 1925
5.10 *Radio News,* December 1924
5.11 Archives Center, NMAH, Smithsonian Institution, Neg. No. 84-9339
6.1, *top left Radio Amateur News,* September 1919
6.1, *bottom left and above Radio News,* December 1920
6.2 *The Talking Machine World,* July 15, 1923
6.3, *above The Radio Age,* May 1923
6.3, *below Radio News,* September 1921

6.4, *above Radio Amateur News*, May
1920
6.4, *left The Wireless Age*, August 1921
6.5, *above Radio News*, August 1920
6.5, *below Radio World*, May 13, 1922
6.6a *The Radio Dealer*, November 1922
6.6b *Radio Merchandising*, May 1923
6.6c *The Radio Dealer*, August 1922
6.6d *The Radio Dealer*, June 1923
6.6e *The Radio Dealer*, August 1922
6.7a *Radio Merchandising*, January
1924
6.7b *Popular Radio*, June 1923
6.7c *Radio Broadcast*, February 1923
6.7d *Radio News*, July 1924
6.7e *Radio Broadcast*, August 1923
6.8 *The Radio Dealer*, August 1923
6.9 Archives Center, NMAH, Smithsonian Institution, Neg. No. 80-20731
6.10, *above Radio Broadcast*, July 1924
6.10, *right Radio Manufacturers of the
1920s*, vol. 2, Alan S. Douglas. Courtesy of Alan S. Douglas and the
DuKane Corporation.
6.11, *above and below* Courtesy of
Zenith Electronics
6.12, *above The Radio Dealer*, February
1924
6.12, *right Radio Age*, April 1924
6.13a *Radio News*, August 1925
6.13b *Radio News*, August 1925
6.13c *The Radio Dealer*, July 1924
6.13d *Radio News*, August 1925
6.13e *Radio News*, August 1925
6.13f *The Radio Dealer*, June 1924
6.13g *Radio News*, June 1924
6.13h *Radio News*, January 1925
6.13i *Radio News*, August 1925
6.13j *Radio News*, March 1925
6.13k *Radio News*, August 1925
6.13l *Radio News*, December 1924
6.13m *Radio Merchandising*, December
1925
6.13n *The Radio Dealer*, November
1924
6.13o *Radio News*, August 1925
6.13p *Popular Radio*, early 1924
6.14, *above* Archives Center, NMAH,
Smithsonian Institution, Neg. No.
90-10152
6.14b, *left The American Magazine*, 1926
6.15a *Radio News*, December 1924
6.15b *Popular Radio*, September 1923
6.15c *Popular Radio*, June 1922
6.15d *Radio News*, early 1921
6.15e *Radio Age*, May 1922

6.16a *Radio Age*, January-February
1923
6.16b *Radio News*, June 1924
6.16c *Radio Manufacturers of the 1920s*,
vol. 2, by Alan Douglas (Collection of
Bruce and Charlotte Mager); courtesy of Alan Douglas
6.16d *The Radio Dealer*, June 1922
6.17 *Radio Broadcast*, October 1923
7.1 *The House Beautiful*, November
1925
7.2 *Radio Broadcast*, September 1927
7.3 Archives Center, NMAH, Smithsonian Institution, Neg. No. 90-10150
7.4c *Radio News*, August 1929
7.5 *Radio News*, April 1928
8.1 General Electric Ad, *The Saturday
Evening Post*, December 20, 1930
8.2 RCA Ad, *The Saturday Evening Post*,
June 17, 1933
8.3 Archives Center, NMAH, Smithsonian Institution, Neg. No. 87-9237
8.4 Archives Center, NMAH, Smithsonian Institution, Neg. No. 90-10151
8.5 Archives Center, NMAH, Smithsonian Institution, Neg. No. 87-16174
8.6a *Radio Retailing*, 1934
8.6b *Radio Retailing*, May 1937
8.6c *Radio Retailing*, November 1936
8.6d *Radio Retailing*, June 1936
8.6e *Radio Retailing*, June 1938
8.7, *above Popular Mechanics*, January
1934
8.8a *Radio-Craft*, November 1938
8.8b *Radio Retailing*, June 1936
8.8c *Radio Retailing*, June 1936
8.8d *Radio Retailing*, May 1936
8.9 *The Saturday Evening Post*, June 17,
1933
8.10a *Modern Mechanix, Hobbies, and
Inventions*, October 1936
8.10b *Modern Mechanix and Inventions*, August 1933
8.10c *Radio News*, October 1937
8.10d *Popular Mechanics*, June 1934
9.1 *Electronics*, August 1938
9.2 *Collier's*, February 18, 1939
9.4, *left Collier's*, November 4, 1939
9.4, *right Saturday Evening Post*, October 17, 1942
9.5 *Collier's*, September 16, 1939
9.6 *Radio News*, March 1940
9.8a *Radio Retailing*, April 1940
9.9a *Radio Retailing*, October 1940
9.9b Zenith brochure
9.9d *Life*, December 8, 1941

9.12a *Collier's*, December 20, 1941
9.12g *Collier's*, December 20, 1941
9.13 Courtesy of Zenith Electronics
9.14, *top Liberty*, June 13, 1942
9.14, *bottom Radio-Craft*, November 1944
10.2 *Holiday*, October 1947
10.3h *Holiday*, April 1947
10.5 *Collier's*, March 4, 1950
10.6 *Collier's*, June 17, 1950
10.8 *Radio and Television Retailing*, September 1947
10.11a Courtesy of Zenith Electronics
11.2, *top Modern Mechanix, Hobbies and Inventions*, January 1937
11.2, *bottom Modern Mechanix, Hobbies and Inventions*, July 1937
11.3, *top Street & Smith's Western Story Magazine*, May 1, 1937
11.3, *center Modern Mechanix, Hobbies and Inventions*, January 1937

11.3, *bottom Street & Smith's Western Story Magazine*, April 3, 1937
11.4, *top Popular Mechanics*, May 1941
11.5, *top right Popular Science*, November 1946
11.5, *bottom left Cavalier*, January 1960
12.2 *Proceedings of the I.R.E. and Waves and Electrons*, vol. 34, no. 4, April 1946
12.3 *Popular Science*, July 1948
12.7 *Science Newsletter*, September 26, 1953
12.8, *left Holiday*, June 1955
13.1 *Holiday*, March 1958
13.2 From a Magnavox Ad, The *Saturday Evening Post*, February 16, 1952
13.5, *left Holiday*, October 1957
13.5, *right Holiday*, March 1958
13.6 *Holiday*, October 1957
13.7, *top Popular Electronics*, October 1955

13.9 *Holiday*, June 1956
13.10, *top to bottom Popular Electronics*, March 1955; *Popular Electronics*, January 1955; *Popular Electronics*, October 1955; *Popular Electronics*, October 1954
13.15 *Holiday*, June 1955
13.16 Publicity photo, courtesy of Zenith Electronics
14.2 *Howard W. Sams, Photofact Folder*, 409–16
14.8 *Senior Scholastic*, September 14, 1960
14.10a *Howard W. Sams, Photofact Folder*, 580–14
14.10b *Howard W. Sams, Photofact Folder*, 588–5
14.10c *Howard W. Sams, Photofact Folder*, 562–5

GENERAL INDEX

About the Author

MICHAEL BRIAN SCHIFFER is Professor of Anthropology and Director of the Laboratory of Traditional Technology at the University of Arizona, where he has taught since 1975. Born in 1947 in Winnipeg, Manitoba, he grew up in Los Angeles. He was educated in anthropology and archaeology at UCLA (B.A., 1969) and the University of Arizona (M.A., 1972; PH.D., 1973). Schiffer has taken part in archaeological field projects in California, Arizona, Arkansas, Chile and Cyprus but is best known for his many and diverse contributions to archaeological method and theory. Especially important have been his writings on behavioral archaeology and on the formation processes of the archaeological record.

Schiffer's current interest is in applying an archaeological perspective to investigating the social and cultural contexts of technology, particularly in the twentieth-century United States. The first book of his that incorporates this interest is *The Portable Radio in American Life*. Presently he is working on additional case studies, including a history of the electric automobile. Schiffer is also the author of *Behavioral Archeology* (1976), *Modern Material Culture Studies: the Archaeology of Us* (1981, edited with Richard A. Gould), *Archaeology* (1982, with William L. Rathje), and *Formation Processes of the Archaeological Record* (1987). In addition, he was founder and editor of the series Advances in Archaeological Method and Theory (1978–1987) and edits the series Archaeological Method and Theory (1989–), published by the University of Arizona Press.

Schiffer and his wife, Annette, live in Tucson, and they have two teen-age sons, Adam and Jeremy.